T0367432

WORLD WAR II
A Pilot's Experience

Robert R. Burch, MD, PILOT

Order this book online at www.trafford.com
or email orders@trafford.com

Most Trafford titles are also available at major online book retailers.

Print information available on the last page.

ISBN: 978-1-4120-3963-5 (sc)

Trafford rev. 06/11/2024

www.trafford.com

North America & international
toll-free: 844-688-6899 (USA & Canada)
fax: 812 355 4082

Table of Contents

Acknowledgements

I had never thought of putting these experiences into print but because of the encouragement of my wife Peggy and 2 children, Robert R. Jr. and Gayle Burch Agnew, I have been able to do so. The journal that I had maintained has been misplaced and lost, so I have relied on records, memory and the input of two crew members, ball turret gunner Al Dawson and radio operator Don Carrick. I am truely appreciative to Al and Don for their assistance. I am thankful to Amy Henry and Robert Burch Jr. for their technical assistance.

My experiences and that of my crew were not unusual but generally typical of the millions of the service men and women of World War II. We were well trained, well equipped, and carried out our assignments as directed. We recognized our purpose and the support and sacrifices of the civilians back home. Relatively few service men of WWII had experienced war before so the adjustment to military life and living under combat conditions was quite large. Humans are very adaptable to change and adjustments and particullarly when a purpose makes it necessary. We all knew the purpose. We did not know the overall scheme or big picture as to how the war would be fought but understood our small assignments and carried them out as best we could. For a war to be fought, soldiers need supplies, equipment, fuel, transportation and communications. Our targets were oil refineries, railroad yards, industry particularly ball bearing plants, bridges and communication centers. We usually heard in preflight briefings, "men destroy your target today because if you don't, then you go back tomorrow".

We, as a crew, became quite close and gave each other much emotional support and confidence. We develop a sense of security when surrounded by men who are capable and reliable. This is something not recognized at the time but it is there and becomes obvious on occasion. I flew with a group of guys who had character and I was grateful for this.

I would like to dedicate this story to my family which has given me continued support, happiness and encouragement.

Peggy, my wife
Robert Jr., my son
Mary Kappel Burch, my daughter in law
Marie Burch, my granddaughter
Amelia Burch, my granddaughter
Gayle Agnew, my daughter
Thomas J. Agnew, my son in law
Thomas A. Agnew, my grandson
Elizabeth Agnew, granddaughter
Annie Agnew, granddaughter

Chapter 1
Introduction Into Military Service

This account is written mostly from memory, so many details may be absent either by choice or forgotten. As an enlisted man, I did not personally receive written orders, so I have no copies for referral. However, later, when I became an officer I did. My career was not exceptional, but ordinary, like most of the 16 million service personnel of World War II. Anyway, for what it is worth, this is as I remember it.

I will begin with December 7, 1941: at that time, I was enrolled as a 17 year old freshman student at Southwestern Louisiana Institute (SLI) in Lafayette, LA (today known as the University of Louisiana at Lafayette). My major was mechanical engineering and I was attending on a football scholarship. Therefore, I was living in the athletic dormitory, which was located in the football stadium. My brother, Warren, was also attending SLI and lived in one of the three agriculture dormitories. Most students attended college on limited funds and therefore did not own a radio or subscribe to a newspaper. Of course, television did not exist at that time. As a result, most young people were not really up on current events. We were aware of the war in Europe and the threat of war with Japan, but not aware of its imminence. The morning of December 7, 1941, I was visiting Warren in his dorm room with his roommates and him when we suddenly heard all this yelling and screaming that "The Japs attacked Pearl Harbor". The students packed into the rooms of those who owned radios while the others were marching up and down the hall carrying brooms as guns and singing patriotic songs. It seemed more like a victory celebration for a major upset over our archrival in football. The significance and consequences of all this had not sunk in fully as yet. This type of reaction spread to the entire campus as news of the attack spread.

President Roosevelt, later that day, gave his famous address announcing the Declaration of War against Japan. This was shortly after Congress had passed the Declaration. By afternoon the students were milling around on the campus and gathering in small groups. This progressed to forming one large group of 300-500 students intent on marching through the streets of Lafayette, ending up downtown on Main Street. Of course, in the past, such a gathering usually ended with students crashing the theaters without tickets and getting to see a movie for free. Dr. Rhiel, Dean of Men at SLI, sensed this could get out of hand because of the high emotions among the students. Warren and I were among the group. Dr. Rhiel met the group as it approached the main gate to the campus, stopped us, and gave an effective speech. He praised everyone's patriotism, emphasized the significance of the event and the insignificance of a downtown march and that the recruiting stations would be open the next morning (Monday, Dec. 8) and if we were really interested in making a contribution, then we should report there. He told us to go back to our dormitories and think about it. With that the group dispersed.

Of course, the war effort immediately became our nation's top priority. Induction stations and recruiting stations opened everywhere. Industry shifted to the manufacture of war material. The famous "Uncle Sam Needs You" sign appeared on just about every street. We were still in the later stages of the depression, and many of the older students had already joined the National Guard to make much needed money. They were called to active duty immediately. These were the first students to leave campus for the military.

In the spring of 1942, recruiters from every branch of service held an open meeting in the campus auditorium. The place was packed with mostly male students, with standing room only. The officer in charge gave a patriotic introduction and emphasized the need for volunteers. He then asked all students interested in the Army to go to one corner of the auditorium, the Navy another, as did the Marines, Air Force and Coast Guard. I joined the Air Force group, and immediately the officer in charge of this group suggested we go outside under a tree in the quadrangle since it was too crowded inside. We were outside only a few minutes when 3 Air Force P-40 airplanes came over at treetop level. What excitement! He signed up just about everyone. Those of us under 18 would need our parent's okay. Physicals and written exams would be held in a few weeks for those desiring to become a pilot, navigator or bombardier. Mother initialy would not sign my papers, but with help from my brother, June, she finally agreed. We were told by the recruiting officer that we probably would be allowed to finish college. I believed this convinced her to sign. A few weeks later I was notified to report for my exams. Both were very extensive; the written lasted one day and the physical was a half day. The results came to us by mail. I was informed that I had passed and would be notified when to report but not before my 18th birthday. The report made no mention of finishing college.

As required by law, I also had to register with my local Draft Board. On June 30, 1942, I registered in Lucy, LA. Albin St. Pierre was the registrar and also the local Post Master.

This was Board One in St. John Parish.

Because of the war effort, jobs became more available. During the summer of 1942, I and some of my friends in Lucy and Edgard became laborers at the Higgins Shipyard on the Industrial Canal in New Orleans. This shipyard was building landing craft and was operational 24 hours a day, seven days a week. We worked the 7 AM to 3 PM shift which meant getting up at 4 AM and leaving our homes in Lucy at 5 AM to drive to the work site. Roger Perret's father ran a truck from Edgard to transport the workers to Higgins, for a fee. The pay was good, $35 for a 40 hour week. This was the hardest work that I have ever done. We worked in the sun all day carrying sheets of steel which were so hot from the sun that, in addition to work gloves, we used folded rags to handle the steel to avoid hand burns. Besides the good pay which I needed for my next year of school, this work made me a better student as I was convinced I was not going to be a laborer for the rest of my life. When I got home in the evening, I was black with rust and sweat.

Later that summer my brother, Donald, learned of an opening as a night watchman at the Calcimine Plant in Norco near his home and he and his wife Mae agreed to my staying with them. I successfully applied for the job and resigned the job with Higgins. The pay was better but I worked from 12 midnight to 8 AM as a night watchman which was not bad. I carried a .38 caliber revolver, fully loaded and in a hip holster. I had shot a similar gun which our famiy owned (unbeknown to Mother) but I am not sure I could have shot someone if the situation arose. I had to make a round of the whole plant every hour and punch a clock placed in strategic locations. When not making rounds, I would load or unload 75 pound sacks of chemicals into or out of box cars. Frequently a hobo would appear asking for food or a place to sleep. Needless to say, I was more scared than he, and I had the gun. Donald and Mae did not charge me room and board, so I was able to save money. But I was glad to return to school.

Saying goodbye to Pat

Boys continued to leave school on a regular basis as they were called into service. Each time a group left, the fraternities held a farewell party. Needless to say, there were many parties. I enlisted officially on December 7, 1942 as a private, serial number 18,171,706, into the Reserve Corps of the United States Army for the duration of the war plus 6 months. This took place in Lafayette, LA. I was allowed to complete that semester. Early in the next semester, I reported to the Induction Center in New Orleans. I and others having to report for duty had our share of parties. I had been dating Patricia Anne Smith (Pat), a girl from Paterson, LA. She, like I, was a sophomore at SLI, and we had been dating for the last year. I mention this because we corresponded regularly for the entire time I was in service.

I left Lafayette and returned to Lucy to visit Mother who was living alone. I was full of enthusiasm up to this point, but, as the day to report approached, my enthusiasm began to change to anxiety. I knew no one who was assigned to report the same day. I was supposed to report at 8 AM with toilet articles and a few personal belongs but no extra clothes. I said my goodbyes to Mother in Lucy and took a bus to New Orleans the day before I was scheduled to report. Mother had arranged for me to spend the night with my brother, June and his wife, Vivian, who were living in New Orleans. The next morning I got up early, and had a good breakfast prepared by Vivian, said goodbye to them and went to the Induction Center by street car. The place was packed with inductees. Here the procedure was to check in, giving your name, and present to the clerk the letter instructing the recruit to report. Each step required getting in line. Next we had to complete a yes or no medical history form; did you ever have pneumonia: yes or no, etc. Next, we lined up for the physical exam. As the line moved, vision was checked by one doctor, then we moved to the next doctor who checked ears, nose and throat, a dentist checked the teeth, next the blood pressure, heart and lungs were checked, next a doctor checked the abdomen and for hernias, and finally the back, joints and for flat feet. This was all done assembly line style. The lab test included chest x-ray, urinalysis, and a blood test. This was my introduction to the Army's method of drawing blood: recruits stood in line, got a needle stuck into a vein, and the blood then dripped into a test tube. About 1 out of 10 fainted, some fainted just watching it done to the guy before him. These were future tough soldiers. The whole process was done while the recruits stood around in shorts. Those who wore no underwear stood around naked. After getting dressed and waiting for the results, I and several others were told to lay down and relax because our blood pressures were too high. If after awhile our blood pressures remained high, we were instructed to go home, rest, refrain from alcohol and caffeine, get a good night's sleep and return the next morning. That's what happened to me. The next day my blood pressure was normal; so I shipped out that day, February 21,1943, by train with a group of other raw recruits. With a sergeant in charge, we traveled by bus to the Texas & Pacific Railroad Station for departure to boot camp training at Sheppard Field in Wichita Falls, TX.

Buck Private Robert R. Burch

Chapter 2
Basic Training At Sheppard Field

Typical of the trains of the time, we traveled in an unairconditioned car pulled by a steam locomotive with coal for fuel. We were a group of about 35-50 recruits. I knew no one but met a few during the trip to Wichita Falls. This was the last time our paths crossed except for one, a recruit named Favrot. I later met up with him in college after the war and at social occasions since. This was an overnight trip, which arrived early the next morning. With anxiety and anticipation still high, we recruits traveled on military buses to Sheppard Field, a boot camp for the Army Air Force. I was on my first military base and now an 18 year old soldier who knew nothing about the military or playing the part of a soldier (a green recruit). Roll call was old hat by this time; one roll call when we prepared to leave the Induction Center, another on boarding the train and still another on leaving the train and arriving at the camp. The train to Wichita Falls was not a troop train but a civilian train, and we had meals in the dining car at government expense.

When we arrived at Sheppard Field, we were told to line up in groups 2 deep and 12 abreast. Our assigned drill sergeant took over. After introducing himself and conducting another roll call, he informed us we no longer belonged to our mothers. He was in charge and in the next seven weeks he was going to make soldiers out of us. No bitching, no crying, we belonged to him. After more of his lecture, we moved to the assigned barracks. The building was one of a large group of identical barracks; one story wood with pointed roof and raised about 2 feet off the ground. The interior was one large room with 24 single bunks, 12 on each side of a center aisle with a foot locker at the foot of the bed. At the front of the building was a room on each side of the aisle. One was the drill sergeant's room and the other for showers, latrine and wash basins. We each hurriedly chose a bed, then had 10 minutes to use the latrine, we were then off to the supply building for clothes. We went single file through a building with a long counter behind which were stacks of clothes on shelves. First we were given a duffle bag for our clothes. The shoes were next. "What size shoes do you wear, soldier?" If the recruit said size 10 you were given a pair of size 10 and told "if they don't fit, then bring them back". This process was repeated with socks, underwear, pants, shirts, coat, over coat, hat, and the works. We returned to the barracks to place our clothes in our footlockers and to hang the appropriate clothes on hangers at the head of the bed. Our sergeant instructed us on how this was to be done in military fashion. This was as it was to be kept and certainly when weekly inspection was conducted by the captain (the company commander). The sergeant explained the barracks bulletin board and told us to read it frequently because all announcements would be posted on it including K P , latrine, and other duty assignments. From the time we arrived to the time we left Shepherd Field, we marched in platoon formation wherever we went as a group; to chow, meetings, training sessions, etc. We were trained an hour or more daily in close order marching; column right, column left, to the rear march, double time, parade rest, and etc. We learned military songs which we sang while marching.

The second day we went to this large auditorium to hear an address by the commander of the base. It was attended by all new recruits, a few thousand. It was sort of like the opening scene in the movie, Patton. It was basically this: "We are at war. Many of you are going to be killed and you are going to kill. Remember your enemy is trained to kill you. You cannot hesitate to kill because it will be either you or him. Your chances of coming home alive will depend on how well you are trained, the care of your weapon, and your discipline. Sheppard Field will teach you all of this." Along with this was a patriotic environment of flags, music, etc. The theme: Kill or Be Killed.

The next lecture was by the field surgeon on basic health habits, foot care, location of company infirmaries and venereal diseases. At the conclusion of this session, we went to our respective infirmaries to have our immunizations brought up to date.

They issued rifles that day but gave us no ammunition. We were tauught how to clean the rifle, take it apart, reassemble it, and shoot it. We learned how to fix a jammed rifle. We spent many hours at the rifle range learning to shoot the M1 rifle. We also learned to shoot a 45 caliber automatic pistol and a Thompson submachine gun. The machine gun was tricky to use because on the automatic setting, when shot, the kick would force the muzzle to move up and right. Therefore one aimed left and below the target, so that in shooting the gun it would swing up and right across the target. We had to demonstrate proficiency in rifle marksmanship but not with the pistol or sub-machine gun. With the rifle, we had to be able to take it apart and reassemble it again in a certain period of time and to make a certain minimum score in marksmanship.

Conditioning with calisthenics, running and the obstacle course was an everyday activity. We were required to be able to do a designated number of sit ups, push ups, run a timed distance, and timed obstacle course . We took frequent short (3-5 mile) hikes with full back packs and rifles, but the biggie was an over night 30 mile hike. This involved full back pack (40-50 pounds), rifle, helmet, and an extra pair of socks. This exercise included digging fox holes and slit trenches, pitching pup tents, as well as doing guard duty. Ambulances and medical personnel were there, too. The usual medical problems were foot blisters, sun burns, and contact dermatitis, as well as the occasional twisted ankle or fracture.

We were trained in the use of the bayonet, hand to hand combat, house to house combat, how to use a compass, and to read a simple map. We had lectures outdoors followed by a demonstration and then group training.

We had training in use of the gas mask and the response to poison gas attacks. Part of this was going through a room filled with tear gas, next was going into a room filed with tear gas wearing a mask, and finally, while doing other duties, having tear gas grenades thrown in the area to test our ability to put on gas masks quickly while continuing our duties.

The company commander (captain), squadron commander (lieutenant) and sergeant major inspected our barracks on Saturday mornings. We had to be in dress uniform, with shoes and belt buckle polished, foot locker open and in order, and everything clean and orderly. At the captain's arrival, the barracks sergeant called us to attention and we remained at attention through out the inspection. The captain noted violations orally, the sergeant major wrote them down and later passed them on in writing to the barracks sergeant. The punishment for violations included: If related to an individual such as unshined shoes or messy foot locker, then he may be assigned to KP or latrine duty. If related to the barracks, then all of us would have to clean the barracks thoroughly or police the area (pick up all trash and cigarette butts on the ground in the area). Late on Saturday mornings was a review of troops on the parade grounds. The troops, were all in full dress, in company formation, and paraded before the review stand and reviewed by the camp commander, his staff and guests. Camp band and the color guard led this parade. It was quiet impressive. Following the parade we were then free for the weekend unless we were on special duty. After the second week, those not on duty received a weekend pass.

There was not very much to do in Wichita Falls. The people on the street were mostly military personnel. We went to the Saturday night USO dance with a big band, many girls, and also food and soft drinks or punch. Soldiers could go up to a girl who was not dancing and request a dance or cut in on a couple dancing and get to dance with that girl. "Cutting in" on a dancing couple was the acceptable practice at that time. We were not allowed to leave the USO with a girl, which was a rule. I am sure that rule was in place to convince the parents to allow their daughters to attend the dance. But of course it gave us an opportunity to meet girls.

Periodically at 2-3 AM everyone was awakened for a "short arm exam". Every one had to stand at the foot of their beds and drop their shorts. A doctor and corpsman would go down the line and on approaching each person, he had to milk his penis so the doctor could check for the presence of urethral discharge, the sign of probable gonorrhea. If positive, the soldier was ordered to report to sick call in the morning for treatment. This was another lesson in loosing one's modesty. The latrines were all open. There were no stalls for showers, toilets, or lavatories, which were all open. The joke was "don't drop your soap when showering".

Our day began at 5:00 AM. We were awakened by our drill sergeant, whose private room was in one corner of our barracks next to the door. "Rise and shine; on your feet. Fall in outside in 5 minutes." If anyone was not immediately up, the sergeant would tilt his bed and roll him out. Everyone was outside the barracks in 5 minutes, called to attention and the roll call began. The responses to roll call may be "here or hey or hi" or any other sound the recruit cared to make. The sergeant then read off the list of assigned duties for the day, including KP, latrine, barracks clean up, etc. These duties were also used as punishment for minor infractions. These assignments were posted on the barracks bulletin board. After roll call we went back in the barracks, got cleaned up, shaved, more properly dressed, made our beds, and then marched to the mess hall for breakfast. After breakfast we returned to the barracks for about 15 minutes then "fell in" (in formation) to begin our scheduled day's activities. We had scheduled activities until 5 PM followed by the evening meal. The rest of the evening was leisure time for rest, trips to the PX (post exchange), movies or whatever. Lights out was at 9:30 PM.

KP was an experience. We had to report to the mess sergeant by 8:00 AM. He assigned your duties. If you were assigned to peeling potatoes, this you did by the sacks. Soldiers peeled, complained, and cursed for 7 or 8 hours. The eyes of the patatoes had to be removed, consequently the skin was peeled deep enough to include the eyes. A four inch diameter potato ended up 3 inches in diameter. Of course when the mess sergeant came by to check, the peeling was being done properly. Dish washing wasn't much better. It wasn't much fun to wash meal trays and the large cooking pots all day. Probably the worse job was taking out garbage and washing out the garbage cans with soap and water. The remarks of the soldiers going through the chow line was something else. No one ever complimented the cook. On the contrary, it was remarks like "I wonder who vomited this food before it was served to us". The plus side of KP was that you ate well. Fortunately, I had KP only twice.

Latrine duty was not too bad. All the toilets, lavatories, mirrors, and floors had to be cleaned. Three or four recruits were assigned to this duty daily. The best part of this assignment, was that, when finished you were done for the day, provided you looked busy. The trick was that, if you went outside or to the PX, you carried a clip board, paper, and a pencil which gave passers by the impression you were carrying out an assignment.

Once a week every one in the barracks had to police the area around the barracks and street. This was picking up trash, bits of paper, cigarette butts, and, as the sergeant said, "anything not nailed down". A couple of days before shipping out, everyone in the barracks had to clean the barracks thoroughly; scrub the floors, wipe down the walls, the lights, bed frames, and the windows inside and out. This was a 4 hour job with the sergeant cracking the whip.

Three things we quickly learned in the military: One: Never volunteer for anything. Two: Always give the appearance of being busy. Three: Stay out of trouble.

Pay was not great. $ 21 per month plus room, board, clothing, and medical care. Once per month, we stood in line to be paid in cash after signing a receipt. We were all buck privates, the lowest rank and lowest paid in the Army.

I never became very close to anyone at Sheppard Field but was friendly to everyone in my barracks. I was closest to the guys in the beds next to mine. One was from Mississippi and the other from Oklahoma. I don't remember the name of either. Both were not interested in flying.

This was basic military training. We learned discipline, life under military conditions, use of basic firearms, close order drill, how to function as a unit, and self preservation. The soldier's impression of Sheppard Field: "The only place in the world where one could be standing in mud and at the same time get dust in his eyes".

We shipped out of Sheppard Field on April 12, 1943. Believe it or not, I enjoyed every minute of boot camp. Everyday was a new experience for me. Next was College Training Detachment for those who qualified for officer training as a pilot, navigator or bombardier. Sheppard Field was one of a number of Army Air Force Bases spread around the country. Anyone entering the Army Air Force (AAF) was assigned to one of these for basic boot camp training. Those not going to flight officer's training were transferred to other bases such as gunnery school, radio school, aircraft maintenance school, etc. Checking out of Sheppard Field, we turned in our rifles and bayonets and were given dress shoes and a few pair of dress socks.

Chapter 3
Off to Dickinson College

Privates were not given individual transfer orders. Privates are one of a group under the supervision of a sergeant, a few other enlisted men and an officer in overall command. We learned a few days in advance that we were shipping on April 12, 1943. Per war time protocol, no one was told ahead of time.

We packed our duffle bags, climbed into trucks, and rode to a railroad spur on the field. We boarded the train as our names were called and assigned to a certain car. The car was old and hot but clean. It had no heat or air conditioning. This was not a Pullman, just a plain passenger car with every seat occupied. There was the overhead luggage rack, toilet on each end, and a vestibule where we placed our duffle bags, each with our names stenciled on it. After everyone boarded, the train moved out to a marshaling yard in town where much shuffling of cars was done. Finally we were on our way, destination still unknown. This was my first and last experience on a troop train and what an experience it was. It was dirty, hot, no sleep, crowded, terrible food, tiring and boring.

The kitchen consisted of three baggage cars set in the middle of the train with the one in the middle being where the cooking was done. The ones on each side were where the serving was done. At meal time, the soldiers were served one car at a time. We got out our mess kits and went single file through the train to the serving car. There we were served our food and filed back to our car to eat. As we walked through the other cars, the remarks about the food was hilarious and at times almost nauseating. The food was not good but when hungry it was adequate. I am sure the mess sergeant complained about his kitchen facilities as much as we did about his food. After eating, we scraped the remnants of food in a garbage can at one end of the car and rinsed our mess kits in another filled with water. This was our routine three times a day.

The train was old, swayed, and creaked a lot. It jolted severely with each start and stop. You'd think that since we were the soldiers who were going to win the war, we would be given top priority, but no. We always were placed on a side track to let other trains go by. Our train was pulled by a steam locomotive fueled with coal. With time, the car became hot, filled with cigarette smoke, stuffy, and filled with B O. Consequently the windows were open but the downside of this was that the smoke and soot from the locomotive entered the car. From this and the moisture of perspiration, everyone soon was black all over; clothes, face and all exposed body parts. It progressively got worse as the trip went on. We made a few stops in marshaling yards along the way where some cars were detached from our train and attached to others, and some cars from other parts of the country were attached to our train.

Sleeping was almost impossible. Some slept on the luggage rack, others on the duffle bags but most tried sleeping in their seats. With the constant talking, moving about of people and the jolting of the train, sleeping was almost impossible. Amusement was card games, dice, reading, conversation and watching the countryside.

The day before arriving at our destination, we were told we were going to Dickinson College in Carlisle, PA. I had never heard of Dickinson or Carlisle, PA. The train arrived in Carlisle at 3 AM, three and a half days after leaving Wichita Falls, TX. During the trip we had no bath, no change of clothes, and arrived dogged tired. This was my one and only experience on a troop train. It was a great experience but once was enough.

Chapter 4
Dickinson College (College Training Detachment)

We were greeted at the station by the commanding officer, other officers, and noncommissioned officers. After finding our duffle bags we "fell in" and roll call began. Everyone was accounted for, so we marched in formation to the campus carrying our duffle bags. The distance was about 1/ 3 mile. April in PA was still cold particularly at night. On arriving at the dormitory we received our room assignment and were free until 11 AM.

The room was a typical college dorm room, with two double bunks and 4 small desks. My roommates were Ralph Booton from Loveland, CO, Wilmore J. Broussard, Jr. from Alexandria, LA, and Bigelow from a suburb of Chicago. Ralph and I have remained friends to this day. Ralph went on to fly B-24 bombers. Broussard ended up flying B-17s, and he and I later trained in B-17s together at Lockbourne Air Field in Columbus Ohio. I have not seen or heard from him since. Bigelow went on to fly P-38 fighters, was stationed in Foggia, Italy, and we saw each other periodically while in Italy. On arriving in our room, we made our beds, then showered and hit the sack. The next morning we were awakened later then usual. The morning was spent unpacking and straightening out our rooms. Later in the morning we had room inspection by one of the officers and a noncom (non commissioned officer). After lunch we had a meeting in the auditorium, and the commanding office spoke to us. We heard an overview of what to expect and what was expected of us. Our pay would be increased to $50 per month, the pay of a cadet plus $25 flight pay. Our classes would be math, physics, english, navigation, meteorology, aviation and physical ed. Navigation included map reading. In Phys Ed, we did much swimming in the college's indoor pool. The women had use of the pool at certain hours and the men and military at other hours. One of the rules of the pool was to swim nude, with a shower required before entering the pool.

Close order drill as scheduled. We had no KP, latrine duty, etc, only guard duty when each of us was assigned.

We were now under an honor code, which was emphasized. If the honor code was broken, it was grounds for dismissal from the Cadets. This was the beginning of officer training, and we were expected to conduct ourselves accordingly. Room inspection would be held every Saturday morning with an occasional surprise inspection. Marching to classes or the mess hall was not done, we just went on our own. We had Saturday afternoon and Sunday free but could only leave the campus with a written pass, providing we did not have assigned duties. We were given our class schedule, books, and supplies. We were frequently reminded that we could "wash out" (dismissed from cadet training) for unsatisfactory performance or infraction of the honor code.

Dickinson College was beautiful, being one of the oldest colleges in the US, dating back to about 1812. The buildings were stone, and many of them were ivy covered. The campus had many beautiful trees and lovely landscaping. The weather in the spring was great. All classes were taught by civilian professors who were on the Dickinson College faculty and were very capable. Only PE was conducted by the military. The schedule was from 8 AM to 4 PM with 1 hour for lunch and 1 hour for library and study during the day. We studied in the rooms at night and lights out at 10 PM. We were awakened for roll call at 6 AM.

My first flight ever in a Piper J3

As part of our course in aviation, we had flying instructions by a civilian instructor in a Piper J3, which is a 2 seat plane. This was at Kingstown Airport, a small airport nearby with grass runways. We wore parachutes. The flight was very exciting and was my very first airplane ride. This was a practical extension of what we learned in the classroom. The students received ground instructions such as familiarization with the aircraft, pre-flight inspection, use of controls, brakes, throttle and instruments, etc. The student did the preflight check, and placed his hands and feet on the controls during take off and landing, but the instructor did the flying initially. Later the students did take offs and landings. While in the air, the student was taught to fly straight and level, to make turns while maintaining altitude, as well as climbing and descending at a specified number of feet per minute. We also learned to do figure eights, fly "S" patterns, rectangular patterns, and do stalls and spins. We also were instructed in taxiing, parking, and securing the aircraft. This was just the basics. A few students learned then that they did not like flying and dropped out of the Cadets. To me it was great and I knew I wanted to fly.

Our only non educational assignment was guard duty. The guard duty involved cadet guards with rifles, no ammunition, and walking guard duty around the campus. The purpose was two fold: one to teach discipline and the other to prevent cadets from going into town without authorization. Guards were assigned around the dormitory section of the campus with each guard walking a two block beat with his rifle shouldered. He was not allowed to converse with anyone except the adjacent guards or the NCO or the Officer of the Guards. Only military personnel could be challenged. One Friday night, I was on guard duty in front of the fraternity and sorority houses. The game of the coeds was to harass the guards. The girls would try to disrupt the guards, try to get them involved in conversation, offer a date, give him her telephone number and tease him about how this guard duty was winning the war, etc. This was a common game of the coeds. Anyone caught negligent on guard duty could be washed out of the Cadets. Guard duty was from 5 PM to 5 AM with individual guards alternating every two hours.

Carlisle, PA, was a very pretty town with friendly people. Entertainment was a movie, occasional dance on campus, and picnics out at the dam on the edge of town. On week ends, the park at the dam was the hangout for the young people. We could canoe and swim on the lake. It was rare that a cadet had to pay for a beer or drink in a bar. The tab was usually picked by a civilian or an officer from the Medical Training

Base in Carlisle (Carlisle Barracks). Physician inductees were trained at Carlisle Barracks.

One of our past times was pillow fights. The four of us (Broussard, Bigelow, Booton and Burch) were the best and took on any room in the dorm.

Carlisle and Dickinson College were delightful but we were ready to move on. The college, the town and the people were great. On July 23, 1943, we marched to the train depot and boarded a train for transfer to Maxwell Field in Montgomery, Alabama.

Cadet Robert R. Burch

Chapter 5
Cadet Classification Center

Maxwell Field in Montgomery, Alabama was the Eastern Division Army Air Force Cadet Classification Center. All of us here were "Officer Qualified and had a desire to fly". Here it would be decided who would receive further training as a pilot, bombardier or navigator. This period was also known as Preflight Training. This is three months of Officer Training, discipline, extensive mental and physical examinations, classroom work, and physical training but no flying.

We traveled again by troop train to Montgomery. This train was less crowded than our previous one from Texas, and the trip was not as long. The railroad had a spur that went right on to the base, so when we disembarked we were there. As usual, we immediately got in formation and marched to the supply building. We were issued new properly fitting cadet uniforms. We received a dress uniform, casual uniforms, shirts, ties, belt, socks, hats, white underwear to replace the olive drab, and three pairs of shoes (dress shoes, sneakers, and army boots). We then marched to the barracks, got our bunk assignments and put away our clothes. We showered, shaved, put on clean clothes, and were ready to go.

The barracks were two stories with one platoon to each floor, typical army barracks. We had single beds not bunk beds like we had at Dickinson College. The beds were on both sides of a central aisle with the head of the bed against the wall. At the foot of the bed was a foot locker for keeping toilet articles, underwear, personal articles, magazines, etc. Between each bed, attached to the wall, was a shelf for hats and under the shelf was a bar for hanging outer clothes (shirts, pants, coats, etc.). The beds had springs similar to a rollaway bed with a 4 inch thick cotton mattress. Each floor had a common bathroom . Each floor was occupied by one platoon (28 cadets) with a sergeant responsible for supervising and training each platoon. Booton, Bigelow, Broussard, and I, who roomed together at Dickinson College, were assigned to the same platoon. In the entrance hall soft drinks, candy, etc. were available at night and on weekends. On the honor system, we helped ourselves and left appropriate money. The barracks were arranged in quadrangles with a small parade ground in the center.

In our routine, we were awakened at 6:00 AM by taps which was played over the PA system. Everyone got up, made his bed, shaved, straightened his foot locker, got dressed and picked up any trash which may have been lying around. Our sergeant in charge made a quick inspection. Following this, the sergeant yelled "fall in" and we immediately lined up in formation outside the barracks. The very first day the sergeant instructed us as to the "routine". He taught us how to make our beds according to Cadet Standards which was with square corners and the cover so tight that a coin would bounce 12 inches when dropped on the bed from the height of the sergeant's head. If the bounce was not 12 inches, the bed was not properly made, and the cadet got a demerit and had to make the bed over. He taught us how to arrange things in the foot locker, on the shelf, on the hanging rack, and how to polish our shoes and brass belt buckle. In formation outside the barracks we were instructed about dress, and personal grooming such as haircut and combing, being clean shaven, with fingernails cut and clean. Our shoes and belt buckles had to be shined. We received demerits for violations. Following this, we marched to the mess hall. When marching, singing was the routine, with mostly military or patriotic songs. Following breakfast, we went back to the barracks for 15 minutes, then we were off to morning activities. Morning activities ended at noon, then we went back to the barracks. At 12:30 PM, we fell in again and marched to the mess hall for lunch. The same was done at 5:30 for the evening meal. On returning to the barracks following the evening meal we were free with time to study, relax, watch a movie or go to the PX . Taps were at 10:00 PM, with lights out and everyone in bed. We had activities in the morning only on Saturdays which always included an inspection by our superior officer. Also on Saturday mornings we often had a parade on the camp parade ground. On Saturday afternoons and Sundays, we were free and could go into town if we wished, with a pass, of course.

Meals were quite different from any experience I had had. Meal time was used to teach table manners which was part of officer training. We marched to and from the mess hall in formation. We had seating assignments at tables of 10. Tables were rectangular with seating for 10, 4 on each side and 1 on each end. Tables had table clothes, utensils and napkins in place and food was served family style. The person at the head acted as host and this position was rotated to each person. Sergeants, who roamed around the tables of the cadets he was responsible for, supervised and taught the table manners. After all the cadets were standing behind their respective chair, the cadet acting as host said "be seated please" and everyone sat. Everyone had to sit erect, elbows off the table, napkins open on the lap. We were taught the proper selection of utensils and use thereof. The knife when not in use had to be across the back of the plate parallel to the edge of the table. Forks on the plate and utensils once used could not be placed on the table. Everyone ate a "square meal", that is with food on the fork, the fork was raised vertically to the height of the mouth then horizontally to the mouth. We were not allowed to put additional food in the mouth until the previous food was chewed and swallowed. If we wanted an additional helping such as bread or other food, the proper request was, "Does any one care for bread?" If no response then "please pass the bread". The cadet host only could order more food for the table. Conversation was encouraged during meals but had to be carried out quietly. At the conclusion of the meal, everyone rose simultaneously and filed out, fell in formation, and marched back to the barracks. The food was good, and, because of the importance of good vision, carrots in some form was part of lunch and dinner daily. During the meal, the sergeant observed everybody's table manners, behavior, and the cleanliness of the table cloth around everyone's plate. Any violations were brought to the cadet's attention, and after the first few days these violations earned demerits.

Discipline was continuously emphasized from the day we entered the service but even more so here. We learned to carry out orders and learned to give them. We were now entirely on the honor system. No lies, no cheating, no stealing, and had to help others work as a unit. We no longer had KP, latrine duty, etc. Our only disciplinary punishment was walking tours. Five demerits in one week resulted in marching around the center parade ground for 1 hr that following weekend. Each demerit above 5, meant an additional 1 hour. This was marching at military pace of 120 steps per minutes, with eyes and head straight ahead. These were done on Saturday afternoons and Sundays. My one experience with walking tours was the one weekend when my brother June, his wife Vivian and Mother drove up to Montgomery to visit me. I hadn't been home or seen any family members since leaving home for the Air Force in February. I was really looking forward to their visit. The week before I received 12 demerits, some for cursing and others for not having made up my bed properly. The tossed coin did not bounce 12 inches. With 8 hours of tours to walk, I did not get to visit with them at all. I requested a delay in the tours to the next weekend and explained the circumstances and even offered to walk twice that many but was denied. My only contact was they seeing me walking tours from a distance while they were sitting in the car. This was also a sacrifice for them and also required planning as gasoline and tires were rationed.

The demerit system was for minor violations of the training system. There were some major violations of the honor code which were not tolerated of an officer candidate. One such violation was stealing. Anyone found guilty of a major violation was "Drummed out of the Cadet Corps". One such incident occurred while I was at Maxwell Field. A fellow cadet, whom I did not know personally, was reported to have stolen money from the soft drink and candy box. The incident was investigated by the MP, and the evidence was presented to the Cadet Honor Board. The cadet also appeared before the Honor Board. He was found guilty. The case was then reviewed by an Officer Board which also found him guilty. The punishment was an entry in his record and to be "drummed out" of the cadets and transferred to another branch of the Army for reassignment or court marshal depending on the seriousness of the violation. No one hears of a proposed drumming out before hand except for the few in command. The night of the drumming out, all cadets are awakened at 11:00 PM and informed there is to be a drumming out at midnight (always at midnight). Everyone is to be in full dress uniiform and must fall in outside the barracks in 20 minutes. After inspection by the sergeant in charge, all cadets marched in formation to the major parade ground. The band was there in full dress playing military music. After all are assembled, the band played the Air Force song. At its conclusion, the drums rolled and the base commanding officer, over the PA system, announced the cadet's name, what he was found guilty of, that he is being escorted off the base, and his name is never to be mentioned again at Maxwell Air Force Base. As soon as he leaves the main gate of the base the drums stop rolling and the band again plays the Air Force song as we march back to the barracks. This is a very emotional occasion, both sad and impressive. We could not help wondering about the person's future and his embarrassment. The odd thing is that none of us ever discussed the occasion even privately. Discipline was rigid but we were constantly reminded that it was necessary to be an effective fighting force and that it would save one's own life as well as that of our buddies.

All cadets received extensive mental and physical examinations. The mental examinations were written and had much to do with problem solving, compositions, navigation, meteorology, and aviation. The physical examination was intensive. In addition to the routine physical, the eye exam included visual field, retinal examination, color recognition, depth perception, night vision, and time required to adjust to night vision. The hearing test included auditory acuity, tendency to dizziness, and orientation test. Part of the routine was a psychiatric evaluation by a psychiatrist. This interview lasted from 20 minutes to 1 hour, including such topics as one's feeling toward his parents and siblings, relationship with boys and girls, and feeling about the war and killing people. Some were given a Rorschach test which is a psychiatric test in which the patient interprets the nature and meaning of a standard group of inkblot cards. There was much kidding and joking among the cadets about the psychiatric evaluation. We also had dexterity tests. Anyone who passed these test, knew that he would continue in officer training and become either a pilot, navigator, or bombardier unless he screwed up.

Cadet Burch walking tours

Navigation training was all class room work and covered map reading and types of maps. Navigation instructions covered dead reckoning which was visual navigation using the map and visual landmarks seen on the ground that were also printed on the map. The preflight flight planning included the distance, direction, altitude, air speed, wind direction and velocity, and ETA (estimated time of arrival). Also important landmarks along the route had to be noted. Aeronautical maps contained the ground elevation, high obstructions such as high towers, rivers, main highways, airports, landmarks such as water towers, railroad and railroad stations, and cities with the outline of the city outlined on the map just as it would be visualized from the air. The maps also show radio beacons and restricted flying areas. We learned celestial navigation by use of a sexton and plot positions from the sun and stars. We were also taught instrument navigation using radio beams, our radios, and instruments. The work consisted of lectures, work problems and examinations. We learned the Morse Code and had to send and receive a certain number of words per minute.

We learned the principals of weather systems formation, air mass movement, cold fronts, warm fronts, up drafts, down drafts, and the effect of altitude on temperature. We learned to read weather maps, recognize different cloud formations such as cumulus, stratus, cirrus and thunder storms. The lectures covered material such as humidity, dew point, air saturation with rain, ice, or snow formation. We were taught weather conditions which were favorable to icing of the wings or carburetor of an air plane. Most lectures were directed to weather as it relates to flying. Covered was the effect of a mountain on the air mass flowing over it with updraft on the windward side and downdraft on the leeward side.

We had lectures on the principles of flying, the theory of lift, and airplane construction. Demonstrations were given on wing construction, controls such as rudder, elevators and ailerons and engine types, parts and function. We worked on engines and learned how they function. We had lectures on propellers and actually worked on them. We learned the different type of propellers, the effect of pitch on function, and the adverse effect of a non feathered propeller on flight when the engine is not running. The importance of weight distribution in an airplane was stressed, as was the importance of not exceeding weight limits.

We had lectures on altitude flying. The need for oxygen was stressed as were the recognition of the first signs of anoxia. We had lectures on avoiding, recognizing and treating "the bends". Time spent in pressure chambers demonstrated it all first hand. Six cadets entered the pressure chamber with 2 instructors and 6 other cadets. The instructors and the 6 other cadets used oxygen masks but the first 6 cadets did not. There were also instructors observing from the outside through glass panels. One instructor operated the chamber. Each of the cadets without oxygen masks on were given a simple task to repetitively perform such as adding by 2's, writing a simple sentence, or doing simple arithmetic problems such as 2 times 2, etc. Each person was supplied with a clipboard, paper and pencil. The six cadets had an oxygen mask available but were not allowed to use it. The second six cadets were using oxygen. The chamber pressure was slowly changed to that experienced at 20,000 feet. As the pressure was brought to that of 12,000 feet and above the cadets were progressively unable to perform their assigned tasks, although they stated they were fine and doing their task well. Their mental and physical state deteriorated to the point of requiring starting oxygen or they would lose consciousness. Also, the earliest signs of bends were demonstrated by a rapid change of pressure in the chamber. All of this made a strong impression on us and brought home the lectures emphasizing the importance of oxygen safety for yourself and others in the airplane. The rule was all had to have an oxygen mask with oxygen on by 12,000 feet and begin putting it on at 10,000 feet if the airplane was still climbing. It was the pilot's responsibility to instruct the crew to go on oxygen, check with each crew member shortly after and periodically to ensure each crew member's well being. The order could vary but generally was as follows: "Pilot to crew, we are going through 10,000 feet, every one go on oxygen and check in". As each crew member went on oxygen and was sure it was functioning properly he would check in: "Tail gunner to pilot, on oxygen and OK, over". Each crew member would check in similarly. The lectures included emphasis on the decrease in temperatures as well as the decrease in atmospheric pressure as altitude was increased. Thus if it was necessary to bail out a high altitude, it was best to free fall to lower altitude before opening the parachute to avoid prolonged exposure to extreme cold and avoid anoxia.

Another area of training was parachuting. We had lectures and demonstrations on the subject. All parachutes at that time were round and made of silk. Nylon was not developed at the time. There were two types: the backpack which flying personnel used and the chest chute which paratroopers used. We were taught and shown the different parachutes, the component parts such as the rip cord, the leg and shoulder straps, shroud lines and the parachute itself. We watched parachute specialists fold and pack the parachutes. Most injuries when parachuting occur on landing, which may result in a leg fracture, or sprained ankle or knee . Lectures and demonstrations were given on the proper way to land and to get the air out of the parachute after landing. If the terrain allowed, we were supposed to land on our feet facing the direction in which the wind was blowing and take the landing shock by allowing the knees to bend a little on contact with the ground and run in the direction in which the wind is blowing. On impact, the parachutist should pull in on the lower shroud lines in order to deflate the parachute. If this is not done and significant wind is blowing the chute will pull the parachutist along the ground.

We were lectured on how to steer our descent to avoid landing in a hazardous location such as a tree, power line, etc. This was done while descending by pulling slightly on the shroud lines on the side of the parachute in the direction to which you wanted to shift. This was pointed out as potentially dangerous for an inexperienced parachutist, if the pull was too hard or too long, because one risked the chance of losing all air in the chute resulting in a free fall, serious injury and possible death. We spent time on the parachute range. We practiced putting on and taking off the parachute, and landing by jumping off a platform 4 to 5 feet high landing on both feet simultaneously followed by taking a few steps. The final test was a simulated jump off a tower about 100 to 150 feet high. The parachute was preopened and the top of the chute was attached to a guide wire which brought the chute with parachutist down to a specified area. Each cadet went to the top platform and under the supervision of a sergeant, put on the parachute harness and made the jump. One jump was all that was required.

We were taught water survival and the use of a mae west (inflatable life jacket). Lectures and movies taught us how to escape burning oil or other fuel on the water surface. The idea was to swim underwater until away from the burning fuel. If the distance was long and it was necessary to surface for air, we were taught to part the burning fuel with an outward motion of the arms as one approached the surface. This allowed the face and mouth to surface between the burning fuel, get a breath of air then go under again and continue swimming. If we went down at sea via parachute it was most important to avoid the shroud lines of the parachute and to swim away from it before inflating the mae west. We were taught to wait for rescue and avoid exhaustion. Time of survival in the water depended upon water temperature. In the North Atlantic survival was 5-10 minutes. We were also given lectures on land survival. They were similar to the teaching in the Boy Scouts, that is fishing, shelter, trapping small animals, starting a fire, and recognizing edible berries and drinkable water.

Much classroom time was spent on recognizing friendly and enemy ships, airplanes, flags and uniforms. This was done mostly with the use of slides and projectors. The flags and uniforms were easy but the different types of airplanes and ships were numerous and much more difficult. We learned the difference in appearance between the various destroyers, cruisers, battleships, fighters and bombers. We had to learn to recognize each in a flash. A slide was projected, and we had to identify it and name the country. With time and practice the identification had to be done from the "after image". The slide was projected for a fraction af a second and all you really saw was the shadow of the projection (the after image). This was like a game and really required attention and concentration. The necessity of this training was obvious since it would not be very long before we would be in combat and have to distinguish friend from foe. This drill was fun.

Physical conditioning was held 5 afternoons per week. The workouts consisted of calisthenics, push ups, sit ups, chin ups, and running. The running was one mile over open rough terrain. My biggest problem was in reaching the required number of push ups and chin ups, since I never had much arm strength. My other problem was tolerating the afternoon sun. We wore only shorts, socks and tennis shoes plus a baseball cap. After 2-4 days my back was blistered. I asked if I could use a tee shirt but this was denied on the basis that everyone had to be dressed similarly. My back was painful and peeling the entire time there. From here on through cadet training, we were required to maintain conditioning that enabled us to do a certain number of push ups, sit ups, and timed running.

Saturdays were for inspections. We, in full dress uniforms, stood at attention at the foot of our bunks. The officer who was a captain (our company commander), a lieutenant and our barracks sergeant came through to inspect each cadet for posture, clean and pressed uniforms, clean hands and finger nails, polished brass buckle and shoes. They wore white gloves to test the shelves, window sills and floor under the bed for dust. The bunks got the coin test, and they checked the foot lockers for orderly arrangement of the contents. Any violations meant demerits. After the inspections were completed, we had a 15 minute break. Then we all fell in to march to the main parade ground for review by the base commanding officer. The reviewing stand was on one side of the parade ground and there stood the commanding officer, his executive officers, company commanders, and any distinguished guests, civilian or military. The cadets, in company formation, entered the parade ground led by the color guard and the band. The cadets entered at one end of the parade ground, marched to the far side, then to the opposite end, and then the near side in front of the reviewing stand. The color guard and band took up positions in front of the reviewing stand but positioned to allow the cadets to march between the color guard and the reviewing stand. The cadets, in company formation, ended up in positions extending across the parade ground facing the reviewing stand. The band played military and patriotic music the entire time. Then there was a period of silence while the commanding officer addressed the cadets. Following this, the cadets marched on the parade ground again led by the color guard and band. On leaving the parade ground, each company marched back to their respective area and were dismissed. Each company was evaluated for their performance and subject to demerits if substandard. This was a very impressive exercise.

Recreation was limited to Saturday afternoons and Sundays, until 6 PM, provided we had no tours to walk. The base had movies, bowling, horse shoes, and pickup soft ball and volley ball. We could go into "town" (Montgomery) but had to have a special pass to spend Saturday night away from the base. Military buses ran from the base into downtown Montgomery and back on a schedule. Cadets could get on or off the bus anywhere along the route. In town, there were more movies, as well as the USO which was dull unless you liked coffee and donuts or unless a dance was going on. The town also had night clubs. Huntington College for girls and a nursing school were located in Montgomery and we would get dates there. We would call the dormitory and usually the house mother answered the phone. If not then one of the coeds answered. The person calling would introduce himself, explain we were cadets at Maxwell Army Air Force Base, knew no one in Montgomery and wondered if she could arrange a date for Saturday night for himself and 2,3, or 4 friends. The answer was always yes but we had to report to the House Mother on arrival. Of course it was war time and most of the boys were in service so the girls were as anxious to meet boys as we were to meet them. We usually went to a movie or went dancing. Sometimes the date was pretty and fun but other times not so.

The people of Montgomery were extremely nice to the cadets, frequently inviting 2 or 3 to spend the weekend at their home. I had this opportunity along with my friends Bigelow and Booton. The lady, whose name I cannot recall, was a widow and had 2 sons in service. She had a beautiful large house in a very pretty part of Montgomery. We went on Saturday afternoon, spent the night and left in the late morning on Sunday. She served a delicious meal on Saturday night and probably one of the best breakfasts I ever had on Sunday. Where she got all this food we never knew since everything was rationed during the war. I will never forget that she served breakfast on the sun porch. We had coffee, milk, juice, pancakes, eggs, biscuits, butter and jelly. She even had fresh flowers on the table. We had much pleasant conversation. She showed us family pictures, and we showed her our pocket pictures. This was my first time in a home and my first home cooked meal since going on active duty. This was a real treat. Many families in Montgomery did the same thing and made the arrangements thrrough the commanding officer's office.

We completed our training at Maxwell Field and I was notified that I would go into pilot training. Needless to say, I was ecstatic. Also Broussard, Bigelow and Booton would go into pilot training, and we were all assigned to primary training in Helena, Arkansas. This was November, 1943. We shipped out by train.

Chapter 6
Primary Flight Training

Helena, Arkansas

We arrived in Helena, Arkansas by train in early November, 1943, and transferred by bus to Thompson-Robbins Air Base for primary flight training. The base was relatively small but had concrete runways and tarmac as well as a few hangers. Our barracks were typical Army barracks. A sergeant was in charge but primarily just to keep things organized and on schedule, none of the fanfare of Maxwell Field. The mess hall served the meals cafeteria style, and we went to mess on our own. Meals were all served only between certain hours. The food was good but still much carrots for our eyes. The base had a Cadet's Club which provided sandwiches, soft drinks, coffee, ice cream and sodas. It also had a juke box and a reading area with magazines and some books as well as writing tables. It was open 6 PM to 9 PM during the week and 12 noon to 10 PM on weekends. Helena was a small town on the banks of the Mississippi River. The town offered nothing to do except a movie. One Saturday night, two other cadets and I were invited to have dinner with the Mayor and his family. The Mayor and his wife had one daughter who was about our age. On arrival there, we found our Base Commander and his wife were also their dinner guests. After much conversation, we had a delicious dinner and more conversation. After dinner the daughter, whose name I forget, took the cadets on a driving tour of Helena. The evening was most enjoyable. The family was extremely nice, the daughter attractive, and the house large and also pretty. That was my last contact with the family and my last trip into Helena except to my flight instructor's home.

After settling in our barracks, we then went to the supply room to be issued our flight clothes. This included flight suits which are similar to the jump suits available today. They are one piece, zippered up the front, with 2 back and 2 side pockets, a small pocket on each sleeve, 2 chest pockets, and 2 large leg pockets. Other items were a leather fight jacket, a winter flying suit which was 2 piece made of sheep skin with the wool on the inside, a flying helmet with goggles, gloves and last but not least our WHITE SILK SCARF which was ours to keep. All of this we kept in our barracks. Needless to say we felt important.

The airplane we trained in was a Fairchild PT 23 powered by a single 220 horsepower Continental radial engine. The propeller was a fixed pitched 2 blade prop. The plane was fabric covered, had an open cockpit with two tandem seats, and a low wing. The landing gear were fixed and with a tail wheel. No flaps. The instruments were the altimeter, air speed indicator, climb and descent indicator, magnetic compass, RPM indicator, airplane attitude indicator, oil pressure and fuel indicators and battery charge indicators. The plane also had a radio and intercom system.

Cadet Burch and PT 23 Primary Trainer

The main field was Thompson-Robbins Air Base with 3 auxiliary fields which were nothing but big level pastures. The runway at these auxiliary fields was a strip down the middle lined with bushel baskets. The idea was to land between the lines of bushel baskets without hitting any. This is where we learned to land and take off. My first flight was on November 3, 1943, with Joseph Crosby as my instructor. All instructors were civilians who were licensed pilots and instructors but either too old for military service or unable to pass the physical exam. They seemed well qualified, very nice but strict. I had Joseph Crosby as my instructor for the first 10 hours of flying. At that point I soloed. The remaining 60 hours was under Instructor Henry N. Raines. Each instructor had 5 student cadets. About once a week we were invited to the instructor's home at night for 2-3 hours of informal instructions on the principals and technique of flying. The instructor's wife usually served soft drinks or lemonade and cookies. He had model airplanes which we used together with pictures to illustrate the topic of discussion. We covered the principals of aviation, techniques of take offs and landings, stalls, spins, rolls, loops, and effects of wind. He pointed out to each our weak points and how to work on them when up soloing. These sessions were not part of our formal study but something most of the instructors chose to do and was appreciated by the cadets.

We had either classroom work or flying every morning or afternoon. We also had one hour of physical fitness every day except on Saturday and Sunday. Physical fitness included running, pull ups, sit ups, pushups, volleyball, basketball, or softball. Class work included principles of aviation, meteorology, navigation, mechanics of the airplane we were flying, and airplane and ship recognition. We were issued books on each subject and had homework nightly and written tests periodically. The day's activities were completed at 4 PM, and we were free the rest of the evening. Dinner was served between 4:30 and 6:30, with lights out at 10 PM. On Saturday morning we had barracks inspection and on occasion a parade review. After this, we were free for the weekend. Again we could go into Helena if we obtained a pass.

Flying was our real interest. Five cadets were assigned to each instructor, who was a civilian. The instructor had authority over us similar to that of the military. We met our instructor at the flight line each day, and he would describe what he had planned for us that day. He would go over in detail each maneuver we were to learn or practice that day. My first flight in primary flight training was on November 3, 1943. Since we had not yet soloed, we went up, one at a time. My first flight at Thompson-Robbins was 32 minutes. This was November and the weather was cold, so we wore our heavy flying clothes. Instruction began with the basics, the precautions when walking the flight line such as avoiding spinning propellers and taxiing aircraft, and being careful of prop wash as frequently foreign objects such as sand could be blown into the eyes. Next we were taught the "walk around" which is to walk around the aircraft to visually check its condition including evidence of oil, fuel or brake fluid leaks, that the tie down ropes were removed, that the chocks were in place, that the tires were properly inflated, that the control restraints had been removed, and there were no holes or defects in the airplane's skin and no obstructions nearby that could be inadvertently taxied into. The walk around takes 2-3 minutes. Any problems were brought to the attention of the crew chief.

Next, the cadet and instructor (Joseph Crosby) climbed into the airplane by stepping on the trailing edge of the wing near the fuselage, walking up the wing to the cockpit, and stepping in. The parachute was one for which the body of the parachute hangs down to the level of the upper posterior aspect of the thigh. The pilot sits on it like a cushion when in the cockpit. Once seated in the cockpit, the pilot went through another check list. The pilot strapped himself in, tightening both the seat belt and shoulder straps. The controls were put through their range of motion, and the pilot checked the brakes to make sure they were on, and attached the radio cable to the helmet. The radio was simple and primarily for intercom and contact with the control tower. The pilot then signaled the crew chief with a circular motion of the upheld hand with the index finger pointed out that the engine would be started. The crew chief then checked the area of the aircraft to be certain nothing was in the way. That being OK, the crew chief then signaled back in a similar manner that it was OK to start the engine. With all controls in a neutral position, the throttle just cracked (pushed forward just a fraction) the starter switch was pushed and the engine started. What excitement. I was seated in the front cockpit, the instructor was in the back cockpit, and all the time we were in contact on intercom. Each item I checked I would relate to him, such as the oil pressure, fuel gauge and compass. After 30 seconds or so with the engine running smoothly, we contacted the control tower for taxi and take off instructions. Once received, we gave the thumbs up to the crew chief indicating we were ready to taxi. The crew chief removed the chocks blocking the wheels, gave us the thumbs up and away we went.

This type of airplane sits nose up at about 10 degrees, obstructing straight ahead vision when on the ground. For this reason, it was necessary to make slow "S turns" when taxiing. The brakes on these airplanes are part of the rudder controls. The foot rudder controls are flat. Extending the foot against the top of the rudder control applies the brakes. The brake to each wheel is independent of the other and therefore this is used in turning during taxiing. The tail wheel has no brakes. It is not possible to lock the brakes on these airplanes. That is why wheel chocks are placed behind and in front of the wheels whenever the airplane is parked. We finally arrived on the taxi strip at the end of the runway. We stopped, applied the brakes, and checked the magnetos. The magnetos control the magnitude of the spark produced by the spark plugs. There are two for each engine, a left and a right on any propeller driven airplane. To check these, the pilot raced the engine to full throttle and turned the magneto control to the left one, back to both then the right one and finally back to both. If the engine missed, then the engine would not get full take off power, so, we have to return to the flight line in that event. Since our magnetos checked out fine, we contacted the control tower for permission to take off. Given permission, we taxied onto the runway, gave full throttle and started down the runway. I had my hands and feet on the controls but the instructor actually controlled the airplane. Once we took off, the instructor gave me the controls and I flew under his instructions. In this first lesson, I did turns of 90, 180, and 360 degrees while maintaining altitude. Mr. Crosby, my instructor, critiqued my control coordination. He had me change altitudes, headings, and fly rectangles and triangles. All of this was to get a feel for the airplane and its controls. He pointed out the auxiliary practice fields and also how to pick out an emergency landing spot if that would become necessary and how to determine wind direction when aloft. He had gone over all of this on the ground before we walked to the airplane but repetition was good. This was my first flight in an open cockpit airplane, and it was exhilarating. From that lesson on, I did the preflight check with the instructor observing.

From the first to the tenth lesson, we practiced landings and take offs at an auxiliary field. As mentioned previously, the auxiliary fields were just an open level field with the runway simulated by 2 lines of bushel baskets separated by a distance the same as the width of the runways at Thompson-Robbins Field. It was soon obvious why the instructors used this set up, as many cadets when learning to land or take off would go off the "runway" hitting a few of the baskets. When this happened, the cadet would have to taxi to the side where the cadet would have to get out to replace the baskets hit. Since this was a frequent occurrence, many extra baskets were kept at each auxiliary field. I suppose I was lucky since I did not have this problem but did have some close calls. Cadets that repetitively made the same error even after explanation from the instructor were subject to discipline. The discipline was walking tours on the flight line wearing a parachute. Since these were fanny chutes, it would bounce and hit you in the butt with every step taken. The instructors also sometimes made the cadet walk back to the base from an auxiliary field carrying his parachute. Some of the cadets have on occasion became lost and had to ask farmers for directions. Also, unbeknown to the instructor, the cadets would hitch a ride from a local motorist and get out of the car or truck before he got to the base. The instructors probably knew this went on, since nobody could possibly walk that fast. They never asked, since they felt the purpose was accomplished, that being to emphasize to the cadet his mistake. The instructor would always stress that this was not punishment but meant to get the cadet to think about this type error since it could cost him his life or the lives of others. After becoming reasonably proficient, we then began to take off and land at Thompson-Robbins Air Base. As we became even more proficient in our flying, we practiced emergency landings at the auxiliary air fields, learned to do spins, loops, rolls, stalls, flying inverted, half loops and figure eights. All of this was great fun.

My first solo flight was on November 23, 1943. That day's flight began as all had. I met Joseph Crosby, my instructor, and flew dual with him for 1 hour and 15 minutes. We landed at an auxiliary field, and he told me over the intercom to taxi to the side and stop. I had no idea what he was up to. He then told me I was ready to solo and just do as I have been and everything would be fine. He told me to carry out some touch and go landings and take offs (landing then taking off again without stopping) and then return for him. Needless to say I was excited and nervous but at the same time confident. I thought I did pretty well. The instructor thought it was acceptable and he complimented me. Mr. Crosby, as we called the instructor, got back in the airplane, and I flew us back to the base, landed and taxied to the flight line. When we got out of the airplane, he gave me a little pep talk, and informed me that I would now have a different instructor, Mr. Henry Raines. He wished me the best and much success in the future. I had 9 hours and 52 minutes flying time including 25 minutes of solo time up to that point at Thompson-Robbins Air Field.

From that time on, we flew for longer periods but not daily because of ground school. On days we flew, part of the flying was with the instructor but most was solo, in order to practice the maneuvers taught. When soloing, one of the fun but unauthorized things we did was fly in, out, and around the cumulus clouds. This was great but the danger was the possibility of a collision with another airplane. Of course we always looked around to be sure no other airplane was in the vicinity.

We had 3 hours of link training while at Thompson-Robbins. This was the forerunner of today's flight simulator. This was my first experience in a link trainer and it was amazingly realistic. In it, we practiced instrument flying, night flying and emergency and cross wind landings. Few cadets washed out of flying status, and those that did either decided they did not like flying and requested transfer or could not pass the flying test. Occasionally a cadet would "freeze on the controls". That is when making a maneuver, he would do nothing with the controls and be unresponsive to the commands of the instructor. The instructor would have to overpower him on the controls. Obviously he would be washed out but gladly so on his part.

I had a total of 70 hours flying time and 3 hours link time at Thompson-Robbins Air Base and enjoyed every minute of it. It was all new experience and interesting. On January 15, 1944, having completed primary training we prepared for transfer to basic training in Greenville, Mississippi.

Chapter 7
Basic Flight Training

Greenville Air Force Base
Greenville, MS

We traveled by train to Greenville, Mississippi, on January 18, 1944. The routine was familiar. We had roll call on arrival, transferring all cadets to the command of Greenville Army Air Force Base. We traveled with a duffle bag and brought along all our clothes issued at Helena, Arkansas, except our heavy fleece lined flying clothes and our flying helmets. We rode on the military bus to the air base. Bigelow, Booton, Broussard, and I were still together. We marched to our respective barracks after hearing our barracks assignments. The base was much larger and more modern than Thompson-Robbins Air Base in Helena. The barracks were typical army barracks floor plan with a two story wood frame. A sergeant was assigned to each barracks primarily to see that we met our schedule and to keep things organized. We each had a foot locker, clothes rack, shelf, and single iron cot, and there was a common latrine on each floor. We were expected to keep the barracks clean and neat but we had no latrine duty, KP, etc. In contrast to Helena, all personnel involved in training were military. Civilians were employed for menial tasks such as kitchen, housekeeping, etc. All flight and classroom instructors were Army Air Force officers.

Shortly after getting settled, there was a meeting of the newcomers. We were told that the Army Air Force expected us to work hard and learn more about flying. They reminded us that we are still just cadets and could wash out. Their stated purpose was to make good pilots of us and that they did not want to wash anyone out.

Our class work was basically the same but more advanced. Principles of aviation, meteorology, navigation, map reading, principals of aircraft design and airplane and ship recognition were the primary courses. We had homework daily. We also studied the aircraft we would be flying. The class work was most interesting since it was presented with daily application in our flying schedule.

Our physical exercise program was one hour daily Monday through Friday but it was not strenuous, mostly running, pull ups, sit ups and push ups. We also played softball, volleyball, or basketball. Our only requirement was that we had to meet the minimum standard requirement of so many push ups, sit ups and a timed run by the time we left Greenville.

Cadet Burch in a BT 13 Basic Trainer

Our basic training aircraft was a Vultee single engine low wing plane powered by a 450 horsepower Pratt and Whitney engine, the BT 13. The propeller was two blade fixed pitch, and the cockpit was closed, 2 seat, tandem, with a sliding canopy. The landing gear were retractable except for the fixed tail wheel. The elevator and ailerons were controlled by a stick just like the PT 23. It had dual controls (back and front seats). The radio and instrument panel were more elaborate. The plane had wing flaps and it had a flap indicator and control. The radio could be used for instrument flying. It had several more navigational instruments. This plane flew faster, higher and had a faster take off and landing speed than the PT 23. The stall out speed was higher as well. Also, this airplane had a cockpit heating and cooling system. For cooling, outside air was circulated through the cockpit. It was heated by circulating the out side air over the engine before entering the cockpit. The parachute was a backpack type chute. Our flying clothes were lighter and less bulky than the fleece lined clothes required for the open cockpit PT 23. We did not need the helmet and so flew wearing our cadet cap. To enter the cockpit, again we stepped up on the trailing edge of the wing and walked up the wing along the fuselage to the cockpit and climbed in. The canopy was not pulled closed until reaching the end of the runway and ready for take off. On landing, the canopy was opened on leaving the runway. The purpose of the open cockpit when on the ground was as a safety feature for rapid exit in case of an emergency.

My first flight in basic training was on January 22, 1944, with Lt. Price. This was a 40 minute flight to orient me as a new student. I went through the same routine as I learned in primary training. We did the "walk around" before entering the cockpit. We previously had cockpit orientation in ground school and so I was familiar with the instrument panel and controls. I went through the same routine once in the cockpit as in primary training. I handled the radio for control tower contacts, I did taxiing, and checked the magnetos, all under Lt. Price's monitoring. With his hands and feet on the controls also, I did the take off. Once in the air at the desired altitude, we surveyed the area, then did some turns, climbs and other basic maneuvers, and then we returned to the airport. I flew the landing pattern and landed with his assistance.

I flew with 3 different instructors while at Greenville Air Base, Lts. Price, Johnston, and Hunter. Initially flights were about one hour but as we progressed and began instrument, cross country and night flying we would fly as much as 5 1/2 hours in one day. Flight time is the time actually in the air. On January 30th. after 7 hours of flight time, I soloed. From then on, I did all the flying. The only time an instructor was on board was when we learned something new such as night flying, instrument flying or cross country flying. Periodically an instructor came along to check the cadet's progress. February 9 was my first cross country flight, which was 1 hr 15 minutes. This was following a flight plan drawn up in navigation class. Also the same day I made my first instrument flight (55 minutes). Even though it was day time, the instructor drew a curtain inside the canopy over the student's cockpit, so that the cadet is unable to see outside the cockpit and was forced to fly by instruments. When preparing for instrument flying, we had to realize that the instruments are always right. We could never rely on our personal sensations concerning the plane's attitude. That is never, never fly by the "seat of your pants when under instrument flight conditions". Doing so is an invitation to crash. BELIEVE IN THE INSTRUMENTS. The student pilot had to learn to be comfortable doing so. For instrument flying, the pilot used primarily 5 basic instruments: One: the attitude indicator. This instrument showed the attitude of the airplane, whether its nose is up or down and whether the wings are level or whether one wing is up or down. Two: the air speed indicator. Obviously knowing the correct air speed is important in flying and navigation. Three: The climb indicator. It should read zero if flying level. It shows the rate of climb or decent in feet per minute. It is a sort of a cross check with the attitude indicator but will show subtle changes which the attitude indicator may not reveal. Fourth: The altitude indicator. It should show the desired altitude. Five: Compass. The pilot has to be sure of the direction of flight. The training airplane has only a magnetic compass. The more advanced airplanes had a radio compass. The other instruments monitored proper function of the airplane. These did not have to be continuously observed.

On February 29, 1944, I was introduced to formation flying. We flew in a formation of 3 airplanes with a lead airplane and one on each wing. The lead airplane was in position 1, the airplane on the right wing in position 2 and the airplane on the left wing was in position 3. The airplanes in position 2 & 3 flew such that their wing tips were approximately 10 feet below and outside the respective wing tip of the airplane in position 1. If a tight formation was called for, then these distances were decreased. When flying formation, we had to have the same air speed as the lead airplane, and the pilot watched the lead airplane and flew off of it. Pilots did not watch the instruments except for an occasional glance. When the lead airplane climbed, you climbed, if it turned, you turned too, always maintaining the same position relative to the lead airlane. This was fun but required constant attention. The pilot of the lead airplane had to fly and make maneuvers that would keep the wing airplanes out of trouble. A few years ago, while the Air Force Thunderbirds were attempting a low level loop in a performance, the lead pilot misjudged the tightness of the loop needed at that altitude and flew the formation into the ground. All were killed. When landing, the formation approaches the field at 1000 to 1500 feet altitude flying parallel to the runway for landing. On reaching the end of the runway, the lead pilot gives the command "peel off". With that the number 3 airplane begins a descending 360 degree turn to the left followed by the # 1 airplane and then the # 2 airplane. Each in turn lands on the runway. The first pilot then turns on to a taxi strip when his airplane slows to a speed sufficient to make the turn.

On February 15, I did my first night flying. I had a 2 hour solo night flight but only after a 1 1/2 hour night flight with my instructor. I made a few take offs and landings then flew in the local area the rest of the time. We had a lecture in ground school on how to do this. Prior to the flight, we sat in a ready room wearing red goggles to allow our eyes to adjust to night vision. We were in our flight clothes and ready to go.

Our instructor reviewed what we were to accomplish on the flight. The numbers and indicators on the instruments and on our watches were coated with phosphorus, and the instrument panel had a ultra violet light so the instruments would light up and could be read easily. After going through the same preflight routine as a day flight, I obtained take off instructions and permission from the tower and took off. After a few take off and landings, I flew around the local area. As instructed, I identified landmarks such as rivers, railroad tracks, highways and towns by the outline of the town made by street lights and lights in homes. Towns and cities are shown on aeronautical maps in the same outline as seen from the air. The outline at night shown by the lights is the same as daytime. All airports had a homing beacon and on clear nights could be seen at great distances. These were rotating beacons with a green light on one side and a white light on the other side. This light indicated a lighted runway airport. A white light beacon indicated that the airport was not lighted. When it was time to return to Greenville Army Air Base, I spotted the beacon and followed it back. As I approached the base, I radioed the control tower, gave my identification and asked for permission to land and for landing instructions. The flight went well, and I was elated. Position lights on an airplane are similar to boats, with the red light on the left wing tip, a green light on the right wing tip and a white light on the tail. This universal position of lights allow the pilot to determine the direction of flight of airplanes in the vicinity. The airplane was also equipped with landing lights.

We were taught cross country flying both during the day and at night. We learned to map out our flight plan, noting alternate air fields along the planned flight path. We plotted the distance, magnetic course, fuel requirements, air speed (true and indicated), checked the weather, and calculated the estimated time of arrival. We were taught that, if we were off course, we should never make a correction in heading until our location is definitely determined. At this stage of training, our navigation was VFR (visual flight rules). This means using visual references when flying. VFR rules require a visibility of 3 miles and a ceiling (base of the clouds) of 1000 feet or more. We were introduced to instrument cross country flying but only with the instructor on board.

We also practiced acrobatics which again included controlled stalls, spins, loops, rolls, lazy eights, and chandelles. We practiced glides. In these, the instructor would cut the engine unexpectedly and declare an emergency landing. We had to quickly pick the best emergency landing spot, determine the wind direction, and glide into position to land. We would not land, however, and, within a few hundred feet above ground, the instructor would restart the engine, climb to a safe altitude, and then he would critique the maneuver. This was all fun.

Here are some of the routine landing and take off patterns that we were taught from the very beginning. The control tower controls the airplanes while taxiing, taking off and landing. Take offs and landings are always into the wind. On take off, the pilot starts the roll near the end of the runway, going down the center of the runway for take off. Once the wheels are off the runway, the pilot continues a climbing flight pattern until clear of the airport flight pattern. Once clear, then he could change the flight heading as desired unless he were given different instructions by the control tower. The flight pattern on landing is different. When in the vicinity of the airport, the pilot contacts the tower by radio, gives airplane identification and location with respect to the airport and requests permission to land and landing instructions. The landing pattern is counter clockwise around the airport. The pilot enters the down wind leg flying parallel and to the right of the intended landing runway. At this point, the airplane is now flying in the landing pattern at the prescribed altitude which is generally 800 to 1000 feet. After passing the end of the intended landing runway by a sufficient distance to land on the end of the runway when on the final approach, the pilot turns 90 degrees left toward an extended imaginary line from the runway while maintaining sufficient altitude. When reaching a point where a medium or shallow turn will align the airplane up with the runway, the pilot then turns the airplane to initiate the final approach. The descent to the runway is made by adjusting the throttle setting and the flaps. The landing gear are usually lowered on the down wind leg. This is the general pattern but may vary with type of airplane, wind direction and speed as well as the terrain around the airport.

Social life was nil. We rarely went into Greenville other than to a movie. The town had nothing to offer. The base had a cadet's club and showed movies. The Cadet's Club and Greenville were both dry. We were busy and training was fun, so this was the extent of our recreation.

I had 72 hours of flying time in basic training and enjoyed every minute of it. We were now ready to move on to advanced training and everyone was anxiously awaiting orders. Generally the small guys were sent to fighter school and the larger guys to bomber training. On March 15 the assignments came out and I was assigned to George Field in Lawrenceville, IL, for multi-engine training. Broussard, Booton and I were transferred to George Field (bomber) and Bigelow was going to single engine advanced training (fighter).

Cadet Robert R. Burch

Chapter 8
Advance Flight Training

George Field
Lawrenceville, Illinois

We transferred from Greenville to George Field in Lawrenceville, Illinois, by regular passenger train rather than a troop train. It had two cars just for the cadets. A sergeant and lieutenant were in charge. Our duffle bags were our only luggage. The trip was relatively short, and so the chair cars were not bad. Lawrenceville was a very small town with no train station. So we disembarked in Vincennes, Indiana which is much larger. Vincennes is just across the Illinois- Indiana state line from Lawrenceville. After roll call, we bused to George Field. We arrived March 25, 1944.

George Field was large with three concrete runways, a nice control tower and a large concrete apron. The cadet's barracks were near the flight line which had flying around the clock. It took a few days to become accustomed to the constant noise of airplane engines. The noise then became just an unnoticed part of the environment. The barracks were one story wood frame with double deck bunks. Although we were responsible for making our beds and keep the area clean and neat, we had no other duties except to study and fly. However, the demerit system and walking tours were still in effect. Each person had a desk and chair for study. We had weekly inspection, and, as cadets, we still had to have shinned shoes, polished brass, good manners, and clean finger nails.

All cadets were assigned to a class, a squadron, and a flight. I and the other cadets who arrived with me were in class 44-E which means we would graduate in the fifth month of 1944. I was assigned to Squadron 1, Flight B. Ralph Booton and Wilmore Broussard were assigned to the same flight as me. In the class of 44-E, there were

4 squadrons and 8 flights in all. Each flight had 48 cadets. Our assignments were according to our flights. All cadets in a flight shared a barracks, ate, flew, had ground school, and PE together. Each squadron had 21 officers as instructors in flying and ground school. The officers were 1st. or 2nd. lieutenants. Some ground school instructors were civilians including women.

The food was good and still included carrots. Meals were served cafeteria style. Food was available at any time since we had round the clock flying at George Field.

Ground school was similar but more advanced than basic training. Again we had principles of aviation, navigation including celestial navigation, map reading, airplane and ship identification, meteorology, and morse code. We had to learn the morse code as well as the use of the equipment to send and receive it. We were required to send and receive a certain number of words per minute. We had classes on formation flying and bomb runs. Again Physical Ed was one hour daily except Saturday and Sunday. We had homework every night. Inspection was on Saturday morning. We were then free the rest of the weekend unless we were assigned to fly or some other duty.

Recreation was adequate. The base had a movie theater, bowling alley, and a cadet club which were all very nice. The cadet club sponsored monthly dances. The USO arranged the dance as well as for the girls to attend the dance. Lawrenceville was a small town with nothing to offer other than one movie theater and a few shops. Vincennes, Indiana, was much bigger with more to offer. Most cadets went into Vincennes when going off base. Scheduled military buses ran to and from the base.

Flying was great. We flew twin engine Beechcraft AT 10. It was a four passenger airplane with pilot and co-pilot seats with rear seats for two others. It was powered by two 560 horsepower Lycoming radial engines. It had low bodied wing, flaps, retractable landing gear, and full instruments including all the navigational instruments available at the time. For solo flying, 4 cadets were in the airplane and each flew a certain amount of time as pilot and co-pilot. If an instructor was along, then he flew as co-pilot and the three cadets each took turns as pilot to learn the maneuvers the instructor was teaching. At times we flew only with 2 on board. My first flight as pilot was one hour and fifty minutes on March 31,1944. My first solo was one hour twenty minutes on April 5,1944. This was my 4th flight at George Field. We kept a pilot log of all flying time as pilot or co-pilot. We practiced emergency landings. Our instructors graded us based on our ability to determine the best location for an emergency landing, and handling the airplane without power. The instructor also checked our ability to handle the airplane with just one engine running. He would, without warning, cut one engine and the cadet would have to quickly increase the power in the other engine, feather the prop in the dead engine, retrim the controls and fly while maintaining our altitude and heading.

We flew cross country flights both in the day and at night. We also flew them on instruments as well as using celestial navigation and radio navigation. These were coordinated with the navigation class in ground school and we practiced them in a link trainer. All total I had 10 hours of link trainer time at George Field. I had two long night cross country navigational flights. One was 3 hour and 15 minutes from George Field to Chanute Field to Scott Field and back to George Field. The other was a 3 hour flight from George Field to Indianapolis to Scott Field and back to George Field. For radio navigation we used long range radio beacons which were located near the large airfields. The aeronautical map indicates the distance and direction from the radio beacon to the airport as well as the radio frequency of the beacon. This is how it worked. The beacon sent out a different morse code signal in each quadrant of the compass: 0 to 90 degrees, 90 to 180, 180 to 270, and 270 to 360 degrees. This allowed us to determine our general location relative to the beacon on turning into that beacon. For example, if the airplane was southwest of the beacon, the pilot took a heading of 45 degrees corrected for cross wind. If the pilot was in the 180-270 degree quadrant and heard the 2 dots (dot dot of the morse code) radio signal then the pilot knew he was in the correct quadrant if that is the morse code signal of that quadrant and is flying to the radio tower. If the pilot were flying too far to the left, the signal would change to the signal of the northwest quadrant, and the pilot knew he had to make a heading correction to bring the airplane back to the southwest quadrant. The pilot could determine that he was getting close to the radio tower if the quadrant was narrow. Once over the radio tower, the pilot would then turn in the direction of the air base. The pilot would then contact the control tower, give his position, and ask permission to land. The tower would give the barometric pressure, visibility, ceiling, wind direction and velocity, and state which runway to land on. The flight controller would also state what altitude to fly in the flight pattern. By the time the airplane reached the air base, the airplane was out of the clouds, and the pilot could make a routine visual landing. This system has long been replaced with a better and easier system. Of course, now we have Satellite Positioning System which is a huge advancement. I had one radio navigation cross country flight from George Field to Evansville and back to George Field in a day flight. I also had two others using instrument navigation. My longest cross country navigational flight was 3 hr 15 min. from George Field to Indianapolis to Cincinnati to Louisville and back to George Field. This was a day flight with an instructor on board checking my navigational and cross country skills. We practiced formation flying both in the day and at night. The formation was less tight in night flying.

The link trainer was a remarkable training tool. It was an exact replica of the cockpit of the airplane, and the controls have the same feel. It can simulate instrument and visual conditions of flying and can simulate normal flying conditions as well as emergencies. It also simulates landings and take offs. It was in the link that we were first exposed to emergency conditions such as loss of power, cross wind landings, loss of one engine shortly after take off, etc. These conditions were controlled by the technician outside the link at his control panel. There are link trainers that duplicate every type of airplane. Since our link duplicated the twin engine Beechcraft, this link had both a pilot and co-pilot.

All total, in advance training I had 10 hours link training, 20 hr. 5 min. instrument flying, 31 hr. 40 min. dual day flying, 4 hr. 45 min. night dual flying, 22 hr. 35 min. day solo, and 11 hr. 45 min. night solo. This gave me a total of 70 hr. and 45 min. advanced flying time. I officially completed my advance training on May 19, 1944.

Our graduation ceremony and swearing in as 2nd Lieutenants was held on May 23, 1944. We were issued officer uniforms a few days prior. Unlike before, these were taylored well. We received a dress uniform and some casual uniforms. An officer's first uniform is government issue, but from then on the officer purchased his own. Needless to say we were all on cloud nine. Posted on the bulletin board was a list of our next assignments. I was going to B-17 school at Lockbourne Air Force Base in Columbus, Ohio, as was Broussard, but Booton was assigned to a B-24 school.

For our graduation ceremony held on May 23, 1944. We wore dress cadet uniforms. The ceremony was an exciting event complete with a band. We were sworn in as officers and individually the base commander presented to us our "WINGS" and our 2nd. lieutenant's bar. What a great day. We were the Class of 44-E. We then received our copy of personnel orders No. 20 dated May 15, 1944 from Headquarters Army Air Forces Eastern Flying Training Command, Maxwell Field, Alabama certifying us as pilots and 2nd Lieutenants effective May 23, 1944. We also received a copy of orders discharging us from service as cadets effective May 22,1944, to accept a commission as a 2nd lieutenant. I also received Orders dated May 23, 1944, transferring me from George Field to Lockbourne Air Force Base in Columbus, Ohio. I was to report there on June 4, 1944, for B17 Transitional Training.

2nd. Lt. Robert R. Burch with new Wings

But first, I was given a 10 day leave. I had not had a leave or been home since February 21, 1943. Needless to say, I was anxious to get home to show off my shiny new silver wings and lieutenant's bars. The other Air Force pilot tradition was the "crushed" dress cap. So as soon as we received our uniforms, out came the metal rim which was designed to keep the cap erect and in the designed shape. Crumpling it up gave the impression of an experienced pilot. Actually this practice had a purpose in that the stiff metal rim precluded wearing a radio headset when flying. This was the practical excuse for this practice, so it was not prohibited by the top brass.

I went home by train. I only had a few days in New Orleans before having to start back, but I enjoyed every moment of my visit. Clyde, my brother, was in service. He was a Coast Guardsman and was overseas manning a supply ship in the Pacific. Two of my other brothers, Warren and Donald, were living in Lake Charles, so I did not get to see them. I visited with the remainder of my family, Dorothy, June, Lucille and Mother (My Dad had passed away when I was five years old). It was great to have home cooked meals and mother cooked all my favorites. I visited no one else other than relatives. I had not been living in New Orleans before service and so I had no friends here. I was treated like a king for those few days, and, on June 1, 1944, told everyone goodbye and boarded a train for Columbus, Ohio. As an officer, we no longer used a duffle bag for travel but now had military green folding suitcases. It had a canvas strip with my name and rank stenciled on.

Chapter 9
B-17 Training

Lockbourne Air Force Base
Columbus, Ohio

As ordered, I reported to Lockbourne Air Force Base for B-17 transitional training on June 4, 1944. I arrived in Columbus, Ohio, by train and rode the military bus to the base. Columbus was a big city, about the size of New Orleans. I enjoyed seeing the city some as we drove around for other stops such as the bus station and the USO Clubs.

Lockbourne was a big base located about 20 miles out of town. As the bus approached the field, I could see some B-17s taking off, and they looked humongous. I thought , "Boy, I will be flying one of those babies". I had never seen one before, only pictures. These and the B-24 were the biggest airplanes in service. On arrival, I went to base headquarters to report for duty as ordered. I was introduced to a few people including the base commander, which was customary for new officers reporting for duty. Being an officer was different than being an enlisted man. I was now treated as an individual and not just one of a group. It seemed odd to now have enlisted men salute me or stand when I walked into the room. I was directed to another office where I had to fill out a number of forms and update my personal papers: home address, nearest kin, person to be notified in case of emergency or death, insurance beneficiary, etc. I was given a map of the base and assigned to officer's quarters. I was told to report to the operations officer the next morning. An enlisted man drove me to my quarters in a jeep.

The officers quarters were nice much like a small apartment building. Each "apartment" had 2 small bedrooms with 2 beds in each, a common sitting room, and a bathroom. Each bedroom had 2 lockers, and the sitting room had 4 small desks and chairs as well as a couch and 2 other chairs. It was not fancy but nice and a big improvement over barracks. I shared these quarters with Wilmore Broussard, Glenn Draper and one other guy whose name I don't recall. The officers quarters were conveniently located, as we could walk to the officer's mess, the flight line, and the PX (Post exchange). Officers had certain privileges We were no longer responsible for our quarters, but instead our quarters were maintained for us. We had complete freedom when not on duty. Officers could leave the base as desired. We had no lights out time at night and could use the Officer's Club and other base facilities as we pleased. It was immediately obvious that being an officer had its benefits and was certainly worth working for.

The airplanes here at Lockbourne were B-17G's, a 4 engine, long distance, heavy bomber. It is powered by four 1,200 horsepower Wright Cyclone turbocharged radial engines. The B-17 had a maximum speed 290 miles per hour with a range of 2000 miles carrying a 5000 pound bomb load. Its maximum altitude was 35,600 feet, armed with thirteen 50 caliber machine guns and had a crew of 10. Maximum bomb load was 17,600 lbs., but, depending on the distance of the mission, the bomb load was less. The plane weighed 65,500 pounds fully loaded. It had retractable landing gear including the tail wheel. It had wing flaps, adjustable pitched propellers which could be feathered. However, the cabin was not pressurized. It had all the latest navigational and other aeronautical equipment. The B-17 was designed by Boeing but during WW II they were also built by Douglas and Lockheed. Over 12,000 B-17's were built. The first combat plane lost in WW II was a B-17. It and its crew were shot down by Japanese fighters on their way to Pearl Harbor on December 7, 1941.

The pilots were issued books for ground school which included a manual on the B-17G with pictures of all of its controls. We also had mock ups of the cockpit to allow us to familiarize ourselves with the location of the controls. We were now graduate pilots and had to make the most of our flying time. This was a multi-engine airplane which required a pilot and co-pilot, but usually the instructor and two other pilots were on board who took turns piloting.

My first flight in a B--17 was on June 6, 1944. By coincidence, this was the same day as the invasion of Normandy. This flight was local and lasted 2 1/2 hours. It included take off and landings as well as general flight maneuvers, all this mostly to get the "feel" of the airplane. The next day I flew a 2 1/2 hour day flight with 1 1/4 hours of it flying on instruments. Instrument flying in fair weather was done with a curtain drawn around the pilot's seat so that he could only see the instruments and controls. As mentioned before, instrument flying required that the pilot believe in the instruments, be relaxed, attentive and had to disregard any personal sensation of airplane's attitude since the sensation might not correspond to the instrument readings. The next day, I did night flying, again with 1 1/4 hours on instruments. Instrument flying was not difficult as long as one was attentive to the instruments. We did much instrument flying both in the daytime and at night. The most challenging task was the instrument take offs. With the pilot's seat hooded with a view only of the instruments and controls, I would line the airplane up with the center of the runway. Using only the instruments to maintain this heading, I would initiate the take off run. As per protocol, the co-pilot called out the ground speed. Pilots can sense if the airplane has enough speed for take off because the airplane feels like it is leaving the runway. Once off the ground, the pilot focuses on the airplane's attitude indicator, to bring about a slow climb with the wings level. We worked on the hooded take offs on two separate days for a total of 7 hours and 35 minutes. This also included instrument flying locally for a brief period of time before landing and doing another one. At the time I thought, "why are we doing this since we would never be cleared for take off with zero visibility". After getting into combat, I learned why. On many days in combat, we had to take off in severe fog with essentially zero visibility.

We did much cross country flying, both in the day and at night to practice navigation and instrument flying. We flew to such cities as Chicago, New York, Nashville, Flint, Michigan, and Dayton. The navigation could be by dead reckoning (visual), celestial or radio. I particularly remember one night flight to Nashville. When we got to the plane our instructor told us we were flying to Nashville and back. Draper would be pilot, he, the instructor, would be the co-pilot and Broussard and I were assigned to do the navigation by any method we chose. Since we were in the navigator's compartment and therefore had no access to the radio navigational instruments, it left us only visual and celestial navigational systems to use. On that night, it was partly cloudy above our flying altitude, limiting celestial navigation. Draper, at our request, got the wind velocity and direction at Lockborne from the control tower. Broussard and I were in the nose of the airplane, sitting at the navigator's table with our maps, pencils and navigational instruments. We did all our calculations using our maps and calculator and plotted our heading and ETA (Estimated

Time of Arrival), taking in account the local wind. We passed these on to Draper, the pilot, and we were on our way. We had not been flying very long when an overcast appeared. This was not surprising since our preflight weather check indicated cloudy conditions between Columbus and Nashville. This now limited our navigation to dead reckoning unless we could find a break in the overcast large enough to allow sufficient time to get a celestial fix. About a half hour into the flight, we began to question where we were. Broussard and I became nervous and were frantically looking for a break in the clouds to get a celestial fix or find a town whose identity we could be sure of. A wind shift caused this problem. The instructor was suspicious of our problem since he would periodically get on the intercom and ask, "How are things going? Do you want any heading change?" We would reply, "No, keep the same heading". We had been taught in navigation classes to never change a heading unless we knew our present location. That was our reasoning since we certainly were lost at the time. We finally flew over a city located on the bank of a river, with city lights that were the exact replica of the city on our aeronautical map. We both agreed on our location, and we reploted our heading taking into account the wind change that we calculated. We were about 50 miles off course. Broussard got on the intercom and gave Draper our new heading and ETA. With these changes, we flew right to Nashville right on our ETA. Our next dilemma was whether we should say anything to our instructor, Lt. Bohlman. We had no problem on our way back since we knew the current winds, and, as we approached the Columbus area, we were familiar with the terrain. As usual, after landing the instructor critiqued the exercise. He said, "I want to congratulate you two on fine navigation. I knew we were off course but in spite of my prodding you did not change course until you were sure of your position and brought us in on time". Broussard and I then quit sweating since we thought we had put something over our instructor.

We had another interesting cross country flight. This was to New York State. We flew over or near Niagara Falls, West Point, and New York City and saw all the points of interest. Our instructor was from New York City, and so he was an excellent tour guide. I am sure he had alerted his family as to his fly over. In civilian life, Lt Bohlman was a dentist.

We practiced day formation flying both at low level and at high altitude. We practiced take offs just as we would do on a mission, join the formation, and climb to high altitude. The formation would fly at high altitude and later return to the field, break from formation, and land. We always reviewed these exercises in ground school before carrying them out.

Burch and Instructor Bohlman at B-17 controls

We had many special exercises or maneuvers to practice. One was short runway take offs and landings. Short take offs were done by starting at the very end of the runway, applying full brakes, using more flaps than usual, run the engines up to full throttle, then release the brakes, and begin the take off run. As soon as the airplane felt light and ready to leave the runway, we would pull back on the wheel and leave the runway. On landing we would try to have touch down as close to the end of the runway as possible. As soon as the airplane was on the runway, the pilot called for flaps up to put more weight on the wheels. The pilot then elevated the tail and applied the brakes while balancing the airplane on the front wheels. To learn pin point landings, we practiced attempting to land on the runway numbers at the end of the runway. To add a little fun, we would bet a coca cola as to who could land closest to or on the runway numbers. When the instructor had his turn, we often equalized the skill by reading the air speed on the final approach 2 to 3 miles per hour slower than it actually was thereby causing him to increase the air speed, resulting in a longer landing glide and therefore over-shooting the numbers. He could not understand why he was always buying the cokes. He was such a nice guy that, even if he knew what we were doing, he went along with our game.

We practiced cross wind landings and take offs. All the large airports and air bases had four runways. Each was at 45 degrees of the compass and so a cross wind was never a problem. Overseas was different in that most air bases had one runway and a cross wind was frequent. So, for practice, a runway with a cross wind would be selected. Taking off cross wind was not a problem. The pilot uses the rudders, controlled with the foot pedals, in order to keep the plane going down the center of the runway. However, in an airplane with a large surface stabilizer, a cross wind has more effect and has a tendency to turn the nose into the wind. However, this is still controlled by the rudders. When the airplane leaves the runway, it is carried to some extent in the direction of the wind until sufficient altitude is reached to correct for this drift by changing the heading a little. A cross wind landing is more difficult. On receiving landing instructions from the control tower, the pilot is told the runway number on which to land and whether there is a significant wind and its direction. When in the final approach for landing, pilots can sense the presence of a cross wind since the airplane will be blown off course from the runway by the wind. He corrects for this by realigning with the runway and "crabbing". That is done by taking a heading slightly into the wind so that the airplane continues on a descending path to the runway. As the wheels are about to touch the runway, the pilot gives some down wind rudder to straighten the airplane so it is now parallel to the runway as the wheels touch. Correcting the crabbing too soon causes the airplane to drift off the runway with dire results.

We also practiced various emergency situations such as having to manually crank the flaps or the landing gear down by hand. When flying, the instructor would unexpectedly cut off 1, 2, or 3 engines to determine how we handled such emergencies and our ability to fly with the engines out. When an engine went out for what ever reason, the throttle to that engine was pulled back, the fuel to the engine was cut off, and the propeller was feathered. A wind milling propeller causes drag. We some times cut back the power on all engines in order to get a feel for how the B-17 glided without power. All of these procedures were covered in ground school, and we also practiced them in the link trainer. We also practiced putting out engine fires. There was a fire extinguisher inside the cowling (engine cover) on each side of each engine with the controls for the fire extinguishers in the cockpit on the instrument panel. In practice we did not release the fire extinguisher but only called out that it was released. We learned that, if the extinguisher did not put out the fire, then we could try to blow it out by opening the cowling flaps to that engine and then dive the airplane to increase the speed rapidly to enable the increased wind flowing over the engine to blow out the fire.

Ground school provided more advanced courses that we had previously, such as meteorology, navigation, aviation, and airplane and ship recognition. We studied the B-17 on classroom mock ups of the engines, controls, instruments, and propellers. We had field mock ups where we learned ditching and which escape hatches to use in those situations. We also learned the escape techniques for parachuting out and for fire after landing. We studied the Norden Bomb Sight and practiced with it using a mock up in a hanger. A bombardier's position was set up on a 30 foot high tower which moved on wheels and was controlled automatically by the bomb sight. On the floor of the hanger was a map of the terrain with an industrial target. The idea was to simulate a bomb run and, using the bomb sight, hit the target. It was fun and more like a game but it did give us an idea of what took place on an actual bomb run.

We had physical education 3 hours a week, and we were required to maintain a certain amount of physical conditioning. The activities were flexible and not regimented, and they were not very difficult.

After flying combat and looking back on the training, I thought it was excellent and I can not remember a single combat experience for which we were not trained.

Free time was ours, and as officers we could leave the base whenever we wanted if we were not on duty. No drinking was allowed the night before flying. Since we flew every day during the week, our nights on the town were Friday and Saturday. Columbus was a nice big city, compared to the towns for the previous bases. Columbus had much to do and lots of girls.

I apparently did well at Lockbourne because on completion of my training I was offered a position there as an instructor. The inticement for this was no combat. But I wanted to fly combat, and so I turned down the offer. Some of my friends thought I was crazy but truthfully I never thought of getting killed in combat. That's youth, I suppose, as I was only 20 years old at the time.

Special order 231 dated August 10, 1944, came out assigning me and others to the

3rd. Air Force Replacement Depot, Plant Park, Tampa, Florida. This was effective August 22, 1944, and we were to report August 31,1944. This meant a 9 day leave. Aubra Draper received the same orders but Broussard was assigned elsewhere.

Draper owned a convertible, and he lived in Amarillo, Texas. He was going to drive home for his leave and invited Broussard and me to ride with him as far as Oklahoma where we could board a train for the rest of the trip home. We figured we could save some money and time. Broussard lived in Alexandria, LA. That trip was a bad experience. We decided to drive continuously so that we would have more time at home. Broussard and Draper did all the driving since I did not know how. It was ironic that I could fly a B-17 but could not drive a car. The tires on the car were all retreads and not in good shape. People could not buy new tires at that time but only retreads, and even those were rationed. Draper's dad had a wholesale food business with many delivery trucks, so he had some tire coupons which he had sent to Draper. Well, between Columbus and Enid, Oklahoma we had 5 blow outs and had to buy more retreads. Fortunately we could find some even though tires were in short supply. Also, we were in uniform and, when shop owners learned we were on our last leave before going overseas, they would go out of their way to find us tires. This was the attitude of everyone in the US toward service men. The other problem on this trip was the convertible top. Draper always wanted it down and I wanted it up. I could not tolerate sun, and this being August, the heat and sun were terrible. We constantly argued about whether the top should be up or down. Car air conditioning did not exist then. We finally compromised, and the top went up in the middle of the day.

On reaching Enid, OK, Draper dropped us off at the railroad station and drove on to Amarillo. Since Oklahoma had many large army bases, the train station was jammed with service personnel and a few civilians. We went in to buy our tickets and after waiting in line, the agent told us that the train would be crowded and the chance of us getting on was slim. The next train would be at the same time the next day. We bought our tickets and laid our plans. Since ours was the next train in, we planned to wait right at the side of the track, and, when the train arrived, we would board through the nearest door. We knew that with all the bases near by, some soldiers would be getting off. Well it wasn't that easy. When the train arrived, it was obvious that it was packed. All the seats were taken, as well as all the spaces in the aisles, vestibules and platforms. When the train stopped, it was like the reverse of people trying to leave a packed burning theater. Everyone rushed to a door. As planned we went to the nearest one, but we were not the first to get there. We finally reached the bottom step but could get no further. The conductor told us that we could not stand there and so had to get off the train. We told him that this was our last leave before going overseas and that we were not getting off the train. Soldiers were pushing and shoving at every door trying to get on. Finally the train began to move. The conductor helped us push so that Broussard and I could reach the platform and he could close the door. It was the same at every stop, and I am sure many of the servicemen were able to spend very little of their leave at home because of the over crowded civilian transportation system. As the train got farther away from large military bases, the train was less crowded. I stood up all the way to Lake Charles.

I wanted to visit with my relatives, Warren, Camille, Donald and Mae in Lake Charles. I had never met Camille since she and Warren were just married, but she was meeting me at the train station. On arrival with many other service men, we had trouble finding each other. When the crowd thinned, a girl who was obviously looking for someone, came up and asked if I was Bobby Burch. Warren sure had married a pretty girl. I spent 2 days and a night in Lake Charles, and then took a train to New Orleans where Dorothy, Lucille, and Mother met me at the station. Dorothy and Lucille both worked, but their companies were good about giving employees a day off for such occasions. I had a very nice leave visiting with family and eating home cooking. I think Mother spent all of her rations cooking my favorite meals. Everything, including food, gasoline, tires, and cigarettes were rationed during World War II. Clothes and cars were not available. I was treated like a king. The leave went by fast and ended with me leaving by train for Tampa, FL.

Chapter 10
Plant Park

Third Air Force
Tampa, Florida

I arrived at Plant Park Replacement Depot on August 31,1944, without incident. Plant Park was a replacement center for the 3rd. Air Force. Air Force flying personnel went there for assignment to combat crews and then were reassigned to a base for combat training. My stay at Plant Park was brief, leaving no time for flying. We were primarily involved in assembling a crew. Our quarters were 10' x 10' tents but sort of fancy for tents. Each tent had a wooden floor about 6 inches off the ground, and the sides were wood up to about 4 feet. The rest of the sides and top were canvas. Each tent had a wooden door. It had a single light bulb hanging from the center of the top. The 6 people in each tent slept in iron frame beds, with 4 inch cotton mattresses on a metal springs. This was the same type of bed we had all through training. Every morning we had roll call followed by a few seemingly pointless meetings. This was only to be sure no one went AWOL.

Volleyball was the common game for exercise and was popular at Plant Park. There was a volleyball league which generated a lot of interest. The second day there, while playing volley ball, a captain called me out of the game. I had no idea why. He introduced himself to me and told me he works in headquarters and is also the coach of the headquarters' volleyball team. He told me they practice everyday at 9 AM. He invited me to join their team. I played a lot of volleyball in high school, and our team had won district championship. So I was considered pretty good. I told him that I would like to but could not since it conflicted with roll call and other scheduled activities. He told me not to worry about that since I would be excused from all activities including roll call as a member of the headquarters team.

I liked that and agreed. I played with the team until I left Pant Park. We had one of the better teams, and we had much fun. I think the rewards were greater later but I was never sure.

Crew assignments came out on September 13, 1944, as special orders 257. We were Crew No. 1619.

Pilot	2nd. Lt. Robert R. Burch	ASN 0833018
Co-Pilot	Flight Officer Robert J. Moran	T63636
Navigator	2nd. Lt. Roger E Sizer	02072371
Engineer	Corporal Archie B. Mills	33656635
Radio Operator	Corporal Donald F. Melley	32649438
Waist Gunner	Corporal Homer A. Walter	15095524
Waist Gunner	Pfc. Floyd S. Matlack	35760899
Ball Turret Gunner	Pfc. Alvin H. Dawson	36899421
Tail Gunner	Pfc. George S. Shirley Jr.	39418863

This was my original crew assignment but, shortly afterwards, headquarters made changes at no request of mine. These were the changes and I was assured that the new assignees were excellent men with good records.

Radio Operator	Cpl. Donald B. Carrick Carrick replaced Melley	34677956
Waist Gunner	Cpl. Tim J. Holly Holly replaced Walter	36903136
Waist Gunner	Cpl. Arthur W. Fouts Fouts replaced Matlack	39421166

Fouts would later become togglier. Bombing technique was changing to pattern bombing. The lead plane of each squadron had a bombardier and the other airplanes in the formation would drop the bombs when he did. There was a toggle switch at the bombardier's position and when the lead plane dropped bombs, the togglier in the other airplanes would flip the toggle switch which would release the bombs. Thus, the name "togglier" for the person with this responsibility. When Fouts became togglier, his position as waist gunner was filled by Nicholas J. Fotos.

Someone from headquarters asked if I would accept Tim Holly as a crew member since the pilot of the crew he was originally assigned to asked that he be replaced because of personality conflicts. He assured me Tim's record was excellent. I agreed, provided I could replace him if the assignment did not work out. That was not necessary because Tim turned out to be an excellent addition to our crew. I had no proof, but I always felt that after my participation with the headquarter's volleyball team and making friends, that my crew assignments were handpicked as some of the best. We had an excellent crew which was well trained, responsible, and could not have performed any better. In addition, they were all very fine guys. I would have put them all up against any in the Air Force. Crew members become very close and are dependent on each other's performance. Our safety, survival, and accomplishment of our assignments depended upon it.

While at Plant Park, I received a letter from Mother stating that Uncle Ed and Aunt Estelle would be visiting their son Billy and would give me a call. Billy was stationed In Tampa, FL, and assigned to 3rd. Air Force headquarters. He was also a member of the 3rd. Air Force band. Uncle Ed called me and invited me to have dinner with them. The band was having a concert so, we attended it and then went to dinner. The concert was great with much stimulating military and patriotic music as well as current favorites. We had an excellent dinner at a lovely restaurant. It was always nice seeing someone from home.

On September 15,1944, we transferred via military vehicle to Drew Field in Tampa, Florida, to begin combat training.

Chapter 11
Drew Field

Tampa, Florida

We met as a crew for the first time on September 15, 1944. No one had met any of the others previously, so we were starting out as complete strangers. I pointed our that we would be together for a long time, and I expected everyone to get along with each other. Any problems of any type, personal or military, should be brought to me rather than just bitching among ourselves. Everyone was expected to do his assignment just as he was trained to do. We then had an open discussion. I informed the crew that we would transfer to Drew Field by military bus later that day and that we were assigned to Squadron S, 327 Base Unit, Drew Field, 3rd. Air Force.

On arrival at Drew Field, a B-17 base, I checked in the crew at headquarters, was given our living quarters assignments and our schedule for the next day. The officer's quarters and enlisted men's quarters were in separate areas of the base. After arriving at Drew, I learned that Lt. William C. Manion, ASN 02065892, a bombardier, would be assigned to our crew for training purposes. The officer's quarters were similar to those at Lockbourne. The units each had an entrance room with a central pot belly stove for heating, 2 lounge chairs and a couch. Off of this room were two bedrooms each with 2 double bunk beds and a common bathroom. One bedroom was occupied by Draper and the officers in his crew and the other by Moran, Sizer, Manion, and me. We had a cleaning service and someone came by early in the morning to light a fire in the pot bellied stove. Our only obligation was making up our beds. The base had good food, and served cafeteria style. We were free to leave the base whenever we were not on duty.

The instructors at Drew Field were former combat aviators, and so we received many pearls about conduct in combat and other interesting stories. The instruction related to combat experiences and flying at altitude. Much of the ground school was a review of previous courses as they relate to combat conditions. We had lectures and practice, teaching us to ditch at sea, escaping the airplane on crash landings, and parachuting from a disabled airplane at high altitude. There are 3 problems in parachuting from high altitude. These are temperature, anoxia (oxygen deprivation), and being a target for enemy fighters. Most of our missions were at altitudes of 30,000 feet. The temperatures at that altitude are very cold, often 30 - 60 degrees below zero or more depending on specific weather conditions of the day. The temperature drops about 2 degrees centigrade and about 3 degrees Fahrenheit for every 1000 feet increase in altitude. The sooner the parachute opens on leaving the airplane, then the longer is the time of exposure to these temperatures, increasing the chance of frostbite and hypothermia. The crews need supplemental oxygen at these altitudes. The airplane oxygen system supplies oxygen to the crew, but parachuting means doing without. Oxygen deprivation occurs above 12,000 feet of altitude causing anoxia and possible syncope (loss of consciousness). As the aviator descended below 12,000 feet, recovery would occur but a completely clear mind may not occur prior to hitting the ground.

The third problem is being a target for enemy fighter pilots. The unwritten code of ethics in aviation combat was never to shoot at an enemy aviator while he is parachuting to safety. This however did not apply to ground troops. This code was often not followed by German fighter pilots for whatever reason, and allied aviators were aware of this. This was not a German policy, but this was an individual pilot reaction. Normally, we were taught when parachuting from lower altitudes to jump from the plane, count 1001,1002, 1003 and then pull the rip cord on the parachute. This gives about 3 seconds to clear the airplane and avoid the open parachute from getting caught on the airplane. Jumping from high altitude is different. We were taught to free fall until we could see the details of objects on the ground and only then pull the rip cord. Free fall from high altitude avoided these three concerns. My crew and I were fortunate in that we were never in a situation that required a bail out, but, in conversation with aviators who did, this is a very scary situation. Aviators try to remember their training, but their main concern is to make sure the parachute opens.

We trained in procedures for emergency exits from the B-17. During an emergency landing, all crew members except the pilot and co-pilot were to go to the waist of the airplane and sit on the floor with their backs against the bulkhead. When the airplane comes to a stop, then they should exit through the waist door. The condition and position of the airplane determined whether the pilot and co-pilot exited from the front or waist of the airplane.

The B-17s that we flew at Drew Field were ones that had been replaced in combat and so were old and not in the best condition. Most were B-17Fs with a few B-17Gs. At that time, radar and the fancy navigational systems and landing and takeoff aids were not yet developed. We practiced simulated combat missions with dud bombs dropped on targets in the Gulf of Mexico and without any bombs but instead using cameras for "bomb" runs on cities. The gunners practiced with live ammunition over targets set up in the Gulf and also against drone airplanes. We worked on take off with grouping into squadron formation then group formation as would be done in combat. In returning to the field after a mission, we practiced breaking from formation just before landing. We did much instrument flying and also much high altitude formation flying just as we would do in combat. Most of these training flights were about 6 hours. We had much time in the link trainers, 16 hours to be exact. Most of this link time was practicing emergencies such as having 1,2, or 3 engines to go out, landing with 1 or 2 engines out, and making a crash landing. This training was essential for combat readiness..

On one night time practice mission, we had one engine to lose its oil pressure. Bob Moran handled this well in feathering the prop, cutting off the fuel to that engine, closing the cowling flaps, and cutting off the electrical power to that engine. We increased the power of the other engines a little, and all was fine. We aborted the mission and headed home. We radioed the Drew Field tower, to let them know we had one engine out, gave them our position, and informed them that we were headed to Drew. We kept the tower informed of our location. As we approached the field, the tower placed all landings and take offs on hold and cleared all flight activity near the field for our emergency landing. As we approached the field, we developed a "run away prop" on another engine. This meant the propeller would not hold its pitch setting and would go into minimum pitch. When in minimum pitch, that engine had no thrust and needed to be feathered. Landing on two engines at night is very difficult. It requires a perfect final approach since it is impossible to gain altitude and fly back around on 2 engines for another landing attempt. One could hold altitude but not gain altitude on 2 engines. Rather than put ourselves in that situation, I told Bob Moran, the co-pilot, to try controlling the pitch using the propeller pitch control located on the instrument panel. Fortunately he was successful. When he got the prop to the proper pitch and released the pitch control, the prop immediately went back to minimum pitch and was in the run away situation again. Thus he had to constantly work this control with one hand while lowering the flaps and landing gear with the other. We had previously instructed the crew to assume positions for an emergency landing and later again notified them that we were approaching the landing. All of the emergency equipment at the field was waiting for us, including the ambulance and fire trucks. The landing went well without incident. Needless to say, we were all happy to be on the ground. At least the pilot and co-pilot could quit sweating.

On another night mission, the weather was questionable before take off but the mission commander decided to go ahead. We were out a couple of hours, and the whole time the weather became progressively worse. The mission was aborted, and all airplanes were instructed to return to base. That meant 25-30 airplanes would be returning at the same time. The problem was that the ceiling was 500 feet with broken clouds to 1000 feet and solid clouds above 1000 feet. Radar, landing aids, and homing aids were not yet available, so the tower controllers had to manage the situation by monitoring the location and altitude of all the airplanes. Each airplane was instructed to contact the tower when it was 10 miles from the field. Each pilot was then instructed to fly a circular pattern around the local navigational radio tower to the left and at an assigned altitude. The altitudes were intervals of 200 feet beginning at 1500 feet. Each airplane would be called in one at a time beginning with the one at the lowest altitude. When that airplane left its altitude, then one at a time each other airplane was instructed to drop 200 feet in altitude. Periodically we were given an update on the altimeter setting, which is the currrent barometric pressure at Drew Field. We were flying on instruments all this time, and this was scary realizing that an airplane you cannot see is somewhere 200 feet above and below you. This is where trusting the instruments pays off. As long as every pillot held his assigned altitude, there was no way 2 airplanes could collide. Finally we were down to 1500 feet and shortly after called in to land and given landing instructions. It was raining hard, with broken clouds down to 400 feet. We were at last on the final approach and completed a safe landing. We called in our location while in the landing pattern and all during the way to completing our landing roll. The weather was so bad that this was the only way the tower could keep track of the airplanes.

One airplane was called in and as it entered and flew the landing pattern, the pilot would call in their position but the tower personel could not see him. There was continued constant communication between the two, with still no recognition of the airplane's location. Finally on the last communication as to location, the pilot informed the control tower that he had landed and was on the end of the runway. Since the control tower personnel still could not see the airplane, they figured out that he had landed at McDill Field instead of Drew Field and told him to switch to the McDill frequency. McDill was a B-26 medium bomber base across Tampa Bay from Drew Field. I learned later from the pilot that, when he switched to the McDill frequency, McDill control tower had been frantically trying to contact him when recognizing the B-17 in their landing pattern. Fortunately, McDill already had its airplanes in and no other airplanes were in the flight pattern. The pilot had an informal hearing, and I am sure he did not get a gold star added to his record. The kidding he received from other pilots was what was most embarrassing. Under the circumstances this was a very understandable error, but we were all glad that it didn't happen to us.

My crew got to know each other well during our stay at Drew Field. The longer we were together, the more I recognized their talents. They were all well trained and very fine guys.

Archie Mills, the engineer, was the oldest of the enlisted men, was more serious and seemed to be the natural leader among the enlised men of the crew. This was fine with me, since he was a responsible person with good judgment. Archie was married, had children, and lived in Ironto, Virginia.

Bob Moran, the co-pilot, an Irishman by descent, was from New Rochelle, NY. He was a happy, fun loving person who did not believe in all this military etiquette. He related well to the enlisted men and was an excellent go between. He and I got along well together, and, during the entire time we flew together, we had only one disagreement. One day after landing and while taxing back to the flight line, we had to cross an active runway. We stopped, and radioed the tower for permission to cross. The control tower told us to hold crossing. Moran misunderstood and thought we were told to cross and argued so. I refused and pulled rank. Shortly thereafter an airplane landed, and we were given permission to cross. He apologized. I accepted and pointed out that, when in doubt, play the safe side. This could have been a serious accident. It illustrates how honest errors can be made.

Roger Sizer, the navigator, from Milwaukee, Wisconsin, and William Manion, the bombardier, from Seattle,Washington, were also married. All the other crew members were single. I could not have had a better crew and knew we would take good care of each other. Lt. Sizer was a family man, with 2 children, and missed his family very much. He talked about them a lot. He made it clear to me in one of our conversations that he wanted to get back to his family after the war and that we should not take any undue risk. I told him we all wanted to go home alive and that I would do my job but for him not to get us lost somewhere. Archie became known as "the Old Man". Al Dawson, the ball turret gunner who was very small in stature (5' 3", 125 lbs.), became known as "Short Round" which was G.I. parlance for a small bullet. Al was from Detroit, Michigan, and was a small person, a requirement to fit into a ball turret. Donald Carrick was from Badin, N. Carolina. He was a thin, wiry guy, energetic and a thinker. Arthur Fouts was from Florin, California. Arthur met a Florida girl and they were married while he was in Tampa. After the War, I had the opportunity to meet his wife, Betty. Nicholas Fotos joined our crew later as a waist gunner. He was from Annapolis, Maryland. Tim Holly, a waist gunner, was from Chicago, Illinois. George Shirley, the tail gunner, was from Los Banos, California.

Tampa, Florida, was a very nice city, pretty with many nice beaches in the area. However, we were there in the fall of the year and it was too cold to swim at the beach. Tampa had lots of night life, which we enjoyed in our free time on Saturday nights. The Latin area of Tampa was known as Eboor City, which had many nice restaurants and night clubs. This was the first time in my life that I had enough money to afford some extravagances. A lieutenant's pay plus flight pay was not bad when compared to the pay of a buck private or cadet. We usually began Saturday evening with a steak dinner and often followed by going to a club with dancing. Tampa had blue laws with no alcohol served after midnight on Saturday until Monday. One popular place in Eboor City must have had a police contact. About 11:45 PM on Saturday, waiters would come around with coffee cups and pour the drinks in the coffee cups and remove the regular drink glasses. After midnight, all alcoholic drinks were served in coffee cups. Almost every Saturday after midnight, there would be a police raid looking for illegal sale of alcohol. But somehow the owner always had advanced notice of the raid because about 10 minutes before the police arrived, the owner would announce over the public address system that the police were on their way and for everyone to drink up. The waiters would check every table to see that the cups were empty and would pour a little coffee in a few cups at most tables. The police would rush in, look around and leave. The only way this could occur successfully was with a police informant.

One Saturday night Moran and I stayed in and hit the sack early, since we were scheduled to fly on Sunday. One cold night, I was sleeping in my long underwear while Moran, as usual, was sleeping in the nude. About 3: 00 AM we suddenly heard a loud voice in our room saying, "OK, you officers on your feet. This is an inspection". I opened my eyes as he turned on the lights and there standing in the doorway was a full Colonel. I thought "What the hell did we do?". Again he shouted "I said get out of bed and I mean now." I looked at Bob Moran, and he was perfectly still acting asleep. At this time of the morning, the pot belly stove heater had burned out and the room was cold. I hopped out of the bed, but Moran lay motionless. The Colonel asked me the name of the other officer. I said "Flight Officer Robert Moran, Sir." He turned to Moran, shook his bed, and yelled "Moran, out of bed immediately or you will wish you were never in this Army." With that Moran hopped out of bed completely nude. We were standing at the side of our beds cold as can be and Moran shivering from the cold still wondering what's going on. The colonel then says, stand at attention. Don't you salute when a superior officer walks in the room?" We both saluted. He then got on Moran for being nude. "Can't you afford to buy clothes? You must gamble. You lost them gambling". Moran responds, "No sir, I don't gamble. I always sleep this way". The colonel responds, "Don't lie to me, I know you gamble". With that, we heard laughter from the next room where Draper was. Then the colonel started laughing and yelled "relax." By that time Moran and I begin to laugh at each other and mad as hell at Draper. It turned out the colonel is with the Inspector General's Office in Washington and was in Tampa to inspect Drew Field. He and Draper had met at a bar in town and had a few drinks together. While driving back to Drew Field, Draper talked him into playing this joke on us. We teased Moran that the colonel included in his report to Washington that Moran sleeps nude. Glenn Draper was always up to some trick. We all had to admit later that it was funny.

During the last 2 weeks at Drew Field, all personnel going overseas had to have a complete physical and dental exam, all their immunizations brought up to date and their personal records as to next of kin, person to be notified in case of death or missing in action, and beneficiary for life insurance, and list of personal belongings. During my dental exam, the colonel who examined my teeth thought my bite was so bad that he wanted to file down my incisor teeth to give me a better bite. I was given an appointment for the next day. I was skeptical, so that night I called my brother June, a physician, for his opinion. He said definitely no. On the next day, I returned to the clinic and informed the colonel that I decided against having this done. He became furious. I asked to be dismissed and left. This all done and the entire crew passing their physical exams, we were now ready to go win the war.

Special orders number 332 dated November 27, 1944, were issued instructing us to report to Hunter Field in Savannah, Georgia, on December 2, 1944. We were ordered to travel by rail and were assigned to CO Sq. S 302nd AAF Base Unit (Staging Wing). We were at that point considered a trained heavy bomber combat crew and on our way to war.

Front row: L to R: Don Carrick, Tim Holly, Nick Fotos, George Shirley, Al Dawson
Back row: L to R: Archie Mills, Bob Moran, Bob Burch, Bill Manion, Art Fouts
Navigator Roger Sizer not in the picture

Chapter 12
Trip Overseas

We arrived at Hunter Field in Savannah, Georgia, on December 2, 1944, as ordered. This was our first stop of many on our trip overseas. After reporting to the operations officer, we were assigned to our quarters. The officers and enlisted men of the crew were quartered separately as was military protocol. Our duty here was to receive a new B-17G and fly it overseas. The airplane had been assembled there in Savannah. It had never flown, and its instruments were not calibrated. Our airplane was B-17G number 44-8582. What a beauty. It had a shining silver fuselage with the only outside markings being those of the US Army Air Force. We had to check out this new airplane for combat readiness and calibrate all of tthe instruments. Airplane no. 44-8582 was one of the airplanes that had newly available radar to enable bombing in spite of cloud coverage. The radar unit replaced the ball turret, and so we had no ball turret. No one on our crew had been trained in using radar. Therefore we would not be flying this airplane when overseas but rather it would be used as a lead airplane with a bombardier trained on radar use in a bombing run.

The first day we went to the flight line where 25 to 30 B-17Gs were lined up. We found 44-8582 and began giving her the once over. Each crew member went to their respective stations and began a tedious check. Moran, Mills and I began by looking over the report that the ground crew had done. Since that was in order, we then began a visual check of the exterior, wheels, wheel struts, engines, controls, propellers, and even fuel tank covers, as well as evidence of fluid leaks. We then entered the airplane. Moran and I went to the cockpit and began checking everything that could be checked on the ground with the engines off. Even such things as the windshield wipers, seat belts and shoulder straps, and windows were checked. In the meantime Mills was checking the control cables, door and exit latches, and any other things mechanical. Carrick checked his radio equipment. Sizer checked the navigational equipment and Manion checked the bombardier's section. Everything that could be checked with the airplane on the ground with the engines off was checked. With this complete, the crew then gathered in the waist section of the airplane and each gave his report. No problems were found, so we then broke for lunch with instructions for Carrick, Mills, Moran, Sizer, Manion and I to return after lunch for instrument calibration.

We returned to our airplane that afternoon. We were to taxi to a special area of the field designed for instrument calibration. After our usual visual pre-flight check, we started the engines and checked that all of the engine instruments were reading correctly. We were then ready to taxi. We were in the second line of airplanes on the flight line and were free to taxi out. However, I began my turn a little late, and it seemed that if I continued, the wing tip would hit the rudder of the airplane parked in front of ours. I thought, "What could be worse than damaging a brand new airplane that had not flown before". I wasn't going to risk it regardless of the embarrassment. With that, I called for the ground crew with a towing tractor to get me out of my dilemma. This they did, and we were on our way. The area was set up so that the airplane could face any direction of the compass and the airplane's magnetic compass could be checked for accuracy and adjusted if necessary. The altimeter was checked for accuracy. On December 19, 1944, we flew the airplane for the first time, that being a 3 1/4 hour test flight . We checked the bombbay doors and the oxygen system, while the gunners test fired their guns over the Atlantic Ocean. All checked out well, and we returned to base. We were scheduled to fly to Dow Field in Bangor, Maine, that afternoon.

While we were at lunch, the ground crew refueled the aircraft and prepared it for "go". That morning the airbase personnel verified and accounted for all the equipment on the airplane. This was done compartment by compartment . We had all been issued our overseas gear. My issue was as follows:

B4 Bag	Summer gloves	Goggles
Summer helmet	Winter helmet	Flyers bag
Oxygen mask	Summer flying suit	Flying gloves
Intermediate jacket	Interim. trousers	Winter shoes
Vest life preserver	parachute emergency kit	Parachute
Bail out cylinder	B-3 Jacket	Sun glasses
Watch	Arctic sleeping bag	Pneumatic mattress

45 cal. pistol	Chest holster	20 rounds ammunition
D 4 computer	Protractor	

Once all of our personal gear was on board, we took off from Hunter Field on the afternoon of December 19, 1944. Our orders # 65 instructed us to leave for Dow Field on or about December 18, 1944. Our flight orders read as follows:

"Route: Hunter Field, Georgia to Dow Field, Maine via Amber Airway # 7 to Philadelphia, PA then fly Blue Airway # 20 from Philadelphia to Allentown, PA. Give position report to Allentown Radio. Fly Red Airway No. 33 from Allentown, PA to Newburg, NY and give position report to Steward Radio. Fly direct from Newburg, NY to Hartford, Conn., then Amber Airway No. 7 from Hartford to Dow Field, Maine. If instructed by radio to land at Grenier Field, turn off on Blue Airway no. 4 north of Boston."

"Alternate Airports: Fort Dix Army Air Base, New Jersey; Syracuse Army Air Base, New York and Grenier Field, New Hampshire. "

"Communications Contacts: communications contacts enroute will be made with the following CAA Radio Range Stations: Charleston Radio, Florence Radio, Raleigh Radio, Richmond Radio, Washington Radio, Philadelphia Radio, Allentown Radio, Steward Radio, Hartford Radio, Boston Radio, Portland Radio and August Radio."

Our flight from Hunter Field was uneventful, but, as we progressed north, the weather became worse. While still flying along Amber Airway no. 7, we were informed by radio that Dow Field was socked in with a winter storm, and we were instructed to land at our alternate air field, Fort Dix Army Air Base, which we did without difficulty.

Just a few hours after landing, I was informed that we were responsible for guarding our airplane ourselves while there. Archie Mills, the engineer, and Don Carrick, the radio operator, volunteered. Before assuming this duty, they went to the American Red Cross office on the base for a snack and for Archie to call his wife. While there, Don decided to call home. His sister who answered the phone, told him that their Dad had passed away at 11:30 that morning. He had 2 brothers in service, and both were in combat in Europe. This had his dad upset, and, on learning that Don was also leaving for combat, this was more emotional stress than his dad could bear. Don and the family all felt this way. Don always carried the blame for his father's death although this was in no way the fault of Don's but one of the many tragedies of war. Don immediately contacted me and gave me the sad news. After expressing our sympathy on behalf of the crew, I asked him if he would like a leave to attend the funeral. Naturally he said yes but he wanted to remain a member of our crew. He didn't think he could do both. I told him that I would see what I could do. I went to Lt. Colonel Walner, Director of Operations, and explained the problem and the fact that Carrick had 2 brothers in combat and could not attend the funeral. He was the only son who might be able to attend. Don wanted to remain with our crew, as we all did. I offered to make the rest of the trip without a radio operator and said that Don could join us in Italy. Col. Walner told me that we were now under the North Atlantic Transport Command headquartered at Dow Field and the decision would have to come from there. He pointed out that this was an unusual request but would contact Dow recommending that the leave be granted. After a few hours, the request was granted and confirmed by operations orders no. 117 signed by Captain Elmer P. Schmit, Assistant Operations Officer. Carrick was allowed to remain assigned to our crew and could join us in Italy. But they did not allow us to make the remainder of the trip without a radio operator. To solve this problem, Air Transport Command temporarily assigned Corporal Raymond Wander, 36513839, to make the trip with us. Raymond was a radio operator with the Air Transport Command, and his orders stated that, on arrival at our destination, he was to immediately begin his return to Dow Field. This was a trying experience, but it was gratifying to know that those in command had heart and were sympathetic of problems of the individual airman. We all saw Don off, expressed our sympathy to him and his family and told him we would be waiting for his return.

Fort Dix was cold, wet, and windy, with the temperature a few degrees above freezing. The cold was penetrating, and, no matter what clothes we wore, it was still cold. We spent the night of December 19 at Fort Dix, and on December 20,. the weather was improved. Dow was now open so we had a 3 3/4 hour flight to Dow, landing in a heavy snow fall.

Dow Field, Bangor, Maine, was our port of embarkation. Our flight plan took the North Atlantic Route to Gioia, Italy, as part of Project No. 90945-R, APO. No. 16740-BY-8. The weather deteriorated over the next 2 days and again flying was suspended. We had good quarters at Dow Field and access to all the facilities of the base including the Officer's Club. However, we were not allowed to leave the base because any improvement in the weather meant we would be leaving for Newfoundland. On December 23, 1944, the weather improved sufficiently for us to depart, but the weather still made for a difficult 3 hour 35 minute trip to Gander, Newfoundland. We flew most of that time on instruments.

The landscape on Newfoundland was snow covered and beautiful. We landed in a heavy snowfall, which continued. As we arrived, a very strong wind was blowing, and so the tower directed us immediately to our assigned location on the flight line. As soon as we reached the flight line, the ground crew immediately placed the wheel chocks, and, as I cut the engines, they tied down the airplane. I had not experienced flying conditions like these previously. I was informed that the wind and snow was predicted to get worse, and the airplane would be blown around and damaged if not secured. The crew placed an engine cover over each engine to prevent freezing. After the airplane was secured, we gathered our personal baggage and piled into a truck which took us to the operations office. There we reported as customary and learned we would not fly out the next day because of deteriorating weather. From there , we went to our quarters. Since we were all tired, after a nice meal, a hot shower and some "chewing the rag", we went to bed. The storm continued into the next day. Moran and I went to Operations to check on our departure prospects and on the weather forecast. The weather forecast was poor with continued snow. The air base would be closed at least for that day and the next. The snow was waist deep everywhere except on the streets and runways. In spite of all the snow, the temperature did not feel that bad and certainly felt warmer than at Fort Dix. I suppose it was a dry cold.

We later borrowed a jeep and drove out to the flight line to check on our airplane. It was snow covered but secure. Moran and I took turns driving the jeep since neither of us knew how to drive very well. With nothing else to do and the wide open flight line, we figured this was an excellent time and place to practice. We practiced parking, backing and making turns. We thought we did pretty well. When doing turns on the icy surface, we discovered it was much fun to skid around on the ice. This soon became our objective with the challenge to skid around 360 degrees. After this we returned the jeep (intact) then went to visit the rest of the crew.

We were told that our earliest possible departure would be the day after Christmas.

The enlisted men planned an evening at the Enlisted Men's Club and the officers planned the same at the Officer's Club. That night, Christmas Eve, after a nice evening meal we went over to the Officer's Club. The club was nice but not fancy or crowded. We went over to the bar to get a drink but were refused. When we asked why, the bartender said we would have to speak with the manager. The manager told us that the club was for members of the Air Transport Command only and totally supported by them. We reminded the manager that we were on our way to combat and that he was not showing much Christmas spirit. We finally prevailed to the point of his offering to sell us a fifth of bourbon but we would have to drink it elsewhere. We accepted his offer and bought our whiskey along with a few soft drinks and returned to our room. With no ice, we placed the cokes and 7 ups in the snow outside our quarters so the drinks would at least be temporarily cold. After a few drinks, Roger Sizer suggested we should have snacks. Well, that was impossible since nothing was available. Then I remembered that at Drew Field, I received a Christmas fruit cake from Mother. The only problem was that it was in the airplane. With that, the four of us put on our warm flight suits to go get it. We took a "short cut" through 4 feet of snow rather than following the street. We finally reached the airplane, retrieved the fruit cake from my flight bag, and returned to our quarters. The irony of all this was that none of us really liked fruit cake and that is why I still had it. In fact, I never understood why Mother sent it to me in the first place except she probably could not get rationed sugar to make cookies, and she knew the fruit cake would not spoil. We ate every morsel of the fruit cake, finished the bourbon, told many jokes and stories, but not once did anyone mention the possibility of getting killed in combat. Sizer and Manion talked about their wives and children and their plans for Christmas. After many Merry Christmas wishes, we retired for a good night's sleep. The next day we all agreed that the fruit cake was the best ever. We learned that the weather the following day may be good enough to fly and we should be prepared to fly out of Gander.

On December 26, 1944, the snow had stopped but the temperature was below freezing and the wind was about 70 miles per hour. The commander decided to go ahead with departure. This was in the early morning hours, so we would be leaving while it was still dark. The entire field was covered with a thin coating of ice including the flight line, runways, and taxi strips. The crew boarded the airplane while the ground crew attached a heater to blow hot air over each engine. The airplane had already been serviced. After the crew was settled in the airplane, the preflight check was completed and the engines were thawed out, each engine was started in sequence while the arplane was still tied down. The ground crew removed the heater and engine cover from one engine, then turned the prop over once and I would start that engine. We did the same for each engine. After all the engines were running and warming up, I had radio contact with the tower. Because of the strong wind and icy conditions, they told us to check the magnetos there while still tied down. After the magnetos were checked, we were ready for take off. One airplane at a time would be untied, chocks removed, and that airplane taxied directly to the runway with out stopping and took off. Also we made as few turns as possible to avoid sliding off the taxi strips. All the snow and ice had been removed from the airplane earlier that day. We were given clearance, untied and off we went on our taxi roll. I could feel the airplane slide on the ice on occasion but nothing serious. I turned on to the runway, applied full throttle to the 4 engines, and we were on our way to Lagens Air Base on Terciera Island, Azores.

Our take off run was short thanks to the 70 MPH head wind. As we gained altitude we were soon in clouds and flying on instruments. Moran and I flew on instruments for 3 hours before breaking out in the clear. Our total night flying was 5 hours 30 minutes which was the first part of our flight. After flying awhile after day break, I saw a huge line of thunderstorms stretched across our intended flight path. Bob and I discussed the possible best approach to cross these and decided it best to fly over or near the cloud tops. If we are unable to do so, we could more easily fly between the thunderstorms. We told the crew over the intercom of the impending rough weather and told them to go on oxygen since we would be at high altitude attempting to fly over the cloud tops. It was soon apparent that these storms extended probably to 40,000 feet or more. We then opted to fly between the storms and warned the crew of rough flying. I selected the least dark area in the line of storms figuring it was between 2 storms and would be less rough. We were soon in this rough weather with lightning all around us, with the airplane bouncing like a cork as we flew on instruments. The static electricity formed a circle at the tips of the propellers. Soon we hit a terrible down draft with the airplane losing altitude at 2000 feet per minute even though I had the airplane in a climbing position. At that point, other then continuing what we were doing and hoping we would exit the down draft soon, we were at the mercy of the weather.

At this time, a peculiar thing happened to me which I never told anyone before. For a fraction of a second it seemed, an image of my Mother appeared, an image of her from the waist up with a light gray background. I couldn't see anything other than her image. Her image then disappeared as quickly as it appeared. I have thought about that a lot but have never really understood "why". Perhaps it was because, when I was growing up, she was always there for me when I needed help. Fortunately we were at high altitude and were soon out of the downdraft. In another 15-20 minutes we escaped this terrible weather altogether. The rest of the flight was smooth. We descended to a lower altitude and came off of the oxygen.

When we approached Terciera Island we radioed the control tower, gave them our identification and location, and asked them for landing instructions and the altimeter setting. The island was not very large, and the field was on one end. As we circled the field, I noticed the runway ran across the entire width of that tip of Terciera Island. The upwind end of it ended at the crest of a 200 foot cliff which dropped off into the ocean. I made a good landing but, as I made the landing roll, it seemed that the runway ended a short distance ahead and that we would roll off the cliff, even though we did not really use that much runway. Just as I was about to call for an emergency stop, we reached the crest in the runway, and at that point I could see lots of runway remaining. What a sigh of relief since we had enough trouble for one day. Our total flying time from Newfoundland was 8 hours 30 minutes of which 5 1/2 hours was night and 3 hours were on instruments. After we pulled up to the flight line and cut the engines, the ground crew chief asked us if we went through bad weather. I answered we and all the other airplanes did. He then told us that one airplane had sprung both wings and is grounded until repaired. The wing tips were sagging 10-12 inches below normal. As a consequence, all airplanes must be inspected before continuing the trip. This would take 2-3 days. This was not bad because it gave us time to see the island a little. The weather was pleasant, cool, and much nicer than Newfoundland. Wine and champagne were dirt cheap. We all bought champagne to take with us. At $1.50 a bottle, the officers bought 12.

Our next stop was Marrakech, French Morocco. Our airplane was found free of damage so we left Lagens on December 29, 1944. Again we had bad weather. On the 6 1/2 hour flight to Marrakech, 2 hours were by instrument flying. I had brought along the book "War and Peace" to read on the trip over but, with the difficult flying conditions, I read very little. The weather was terrible over the entire Mediterranean region. Therefore, we remained in Marrakech until January 8, 1945. The language in Morocco is French. Having grown up in Edgard, a French speaking community, and having studied two years of French in high school, I could communicate a little with the locals. I became better at it each day. Marrakech was nice and much like it is portrayed in the movies. One day we drove out in the desert to see the Sheik's castle. We were not allowed to enter but viewed it from a distance. The castle and the grounds were beautiful, but beyond that it was all desert.

Most of the Arabs were poor and uneducated. The men wore long ankle length one piece shapeless tunics. Many tunics were just US Army mattress covers with a hole cut for the arms and neck. The Arab women all wore veils. Camels and bicycles were the common means of transportation. I remember an incident at the Officer's Club. As we arrived at the club, the manager, who could not speak Arabic, was instructing an Arab, who could not speak English, to water the outside plants. He had 2 buckets of water and demonstrated by watering some plants with water from one bucket. He then pointed to the plants and water and says "you". The Arab poured the water from the full bucket into the empty bucket the manager had used, then from that bucket poured water on some of the plants. He then went off to get more water and only used that same bucket to pour the water on the plants. We laughed as we went by and the manager turns to us and said "At least he is getting the job done".

On January 8, 1945 we left Marrakech for Tunis, Tunisia, a 4 hour 10 minute flight. On leaving Marrakech, we flew over Oujda in western French Morocco on to Oran, Algeria, located in northern Algeria on the Albora Sea. From Oran we then flew to Algiers, Algeria, and then to Tunis, Tunisia. All this was nonstop. Tunis was our only overnight stop. On our trip over north Africa at high altitude, we could see the Sahara Desert and what a desolate place it is. There was nothing but sand as far as we could see, no buildings, trees, growth of any sort, or water.

The next day, January 9, we were scheduled to fly to our final destination but inclement weather prevented this. We were diverted to Grottaglae in southern Italy. On the way across the Mediterranean, we flew over the eastern edge of Sicily. As we approached Sicily, I could see Mt. Etna ahead and to the west. I asked the crew over the intercom if any of them had flown over a volcano before. None of us had, so I told them we would fly a little detour over Mt. Etna. Thinking this was no big deal, I banked the plane toward Mt. Etna. Just as we got over Mt.Etna, one of the crew came over the intercom saying, "Lieutenant, two spitfire fighter planes are approaching at

6 o'clock". I looked back to see a British spitfire coming up on each wing. Carrick told me that they were trying to make radio contact. I switched the radio frequency to airplane to airplane. This voice comes over, "Yank, where do you think you are going?" I responded, "We are headed to southern Italy but took a little tourist detour to see Mt. Etna." He then responded, "I advise you to get back on course and take your tour after the war." I said "Roger" and got back on course as we gave each other a little salute. At least we accomplished our mission and saw Mt. Etna.

We finally reached Gottaglae after a four hour ten minute flight. It was cold and damp with snow on the ground. At operations they told us that we would have to sleep in our airplane because they had no sleeping accommodations for us. I so informed the crew who responded with much vulgarity. A few hours later, a master sergeant appeared on the scene and offered sleeping facilities for the officers but could not accommodate the rest of the crew. He did state he would try to find sleeping space for the enlisted men. We were hesitant to leave the enlisted men but they insisted we accept the offer. The sergeant informed us that we would be staying with him and his buddies. In the States this would be verboten, but in combat protocol was more relaxed. The facility was great, a limestone building with a kitchen, a large bed room with a fire place, washstand, lights and windows. We were quite surprised and content. They gave each of us a folding army cot on which we placed our Arctic sleeping bags. They even gave us pillows. We initially could not understand how these enlisted men had such luxury when everyone else had tents. We soon found out that these four master sergeants essentially ran the base on a daily basis. One ran the motor pool, another ran the mess hall, the third was the ranking enlisted man in headquarters, and the fourth ran the supply unit. These four were with the first American unit to invade Africa at the start of the war in Europe and had been in combat ever since. They had many interesting stories to tell. Supplies at the start of the war were in short supply and they learned to make do with captured German equipment. They learned to use parts from damaged equipment to keep other less damaged equipment running. Often the food supply was limited and rationing was necessary. Most of their friends who were part of the initial invasion were killed or injured then or in subsequent fighting. In short, these soldiers learned to live by their wits. All of them received field promotions to their present rank, and all subsequently turned down offers of field promotions to officers.

These master sergeants now had it made. They had this great living accommodation, use of a jeep, good food, and an apartment in Gottaglae. They treated us like kings. They served us fresh eggs, juice, bacon, toast and coffee for breakfast. The evening meal included a cocktail before a steak, baked potato and salad, all this served in the room. I would not have believed some of their stories if I had not witnessed these guys' ability to get things done. Their war stories were fascinating. After the first night, these guy arranged sleeping accommodations for our enlisted men. Because of the poor weather conditions throughout most of Italy, we remained there for 10 days. On our departure, we offered these guys something for their hospitality but they refused. We wished each other good luck and departed. Shortly after the war in Europe ended and I was still in Italy, I heard that four master sergeants in southern Italy were court marshalled for misuse of authority and fraud. At the end of the war, many officers requested transfer to combat zones because this was good for one's record. On arrival in the combat zone, it was commonplace for them to enforce protocol and formality to the fullest. Without any combat experience and little understanding of what some of these guys had been through, court marshals were common. It may well be that our friends in Grottaglae were the ones court marshalled and undeservingly so, considering their combat contribution. I will never know if this was due to their past actions or some recent escapade.

On January 19, 1945, we took off for our final destination, Celone Field in Foggia, Italy. However, the runway at Celone Field was being replaced, so the 463rd. had moved its airplanes to the 301st. Bomb Group Field temporarily. The personnel were still at Celone Field, even though they were flying out of the 301st Field. As we approached the field, I contacted the tower. They gave me landing instructions and the altimeter setting, and also told me that I would be landing in a strong cross wind. This airbase had one narrow runway made of steel mesh, made of strips of steel about 1 foot in length, 12 feet across, and about 3/8 inch thick. The steel was perforated, and each strip was hooked together to make the length of the runway. During rain, the mud would rise through the holes and make the runway slick. As I approached for landing and correcting for cross wind, I straightened out too soon and was going to land with one wheel off the runway. Recognizing this, I gave full power, pulled up and flew around and made a second landing attempt which was a good one. By radio, we were directed to our revetment and parked. After 32 days of travel time, we finally arrived at our combat destination.

Chapter 13
Combat

A driver with a truck met us at our revetment on arrival at the 301st Bomb Group Field. Since we were actually assigned to the 463rd. Bomb Group, 15th Air Force, at the next airfield,we had to be driven there. We collected our gear, threw it into the back of the truck, climbed in and drove off to Celone Field, the home of the 463rd. Bomb Group which was also at Foggia. The ride was about 30 or 40 minutes. On arrival at group headquarters, I was supposed to hop out of the truck and report our arrival. Out the back end of the truck was nothing but mud and water. I yelled to the driver to pull up out of the mud. He yelled back that I had better get used to it because the entire field is this way. He pulled up a little, but it was no better. I hopped out and immediately my shiny boots were no longer shiny. I checked us all in, reported where the airlane was located, and told them what was happening with Don Carrick and Cpl. Raymond Wander.

Our crew was assigned to the 775th. Squadron, Allyn's Irish Orphans, named after the first squadron commander, Robert H. Allyn. The 463 Bomb Group was nicknamed the Swoose Group by its first Commander, Lt. Colonel Frank Kurtz. During the entire war up to that point, Col. Kurtz had flown a B-17 named the Swoose. At the start of WW II that B-17 was in the Pacific.. Because of the shortage of spare parts, airplane repairs were made using parts of airplanes that were more severely damaged. The Swoose was so named because it contained so many scavenged parts that it was half swan and half goose or a "Swoose". The Swoose later became the only combat airplane to fly during the entire war. After being replaced in combat, it became General George Brett's command airplane in Austrailia and later Central and South America. It also toured the United States selling war bonds. It now belongs to the Smithsonian Institute and is being restored. Colonel Kurtz was so fond of the name that he later named his new born daughter "Swoosie". Swoosie grew up to be a successful movie actress. The squadron commander at the time of our arrival was

Major Jack A. Mendell. We then rode to the orderly room for the 775th. squadron, which was the squadron headquarters. The building was small, with only 2 rooms, built of limestone blocks. The smaller of the 2 rooms was the office of the squadron commander, while the other was the general office and was much larger. Outside of the building was a bulletin board where important announcements such as the flight schedule for the next day were posted. Our crew met the squadron commander, we completed our paper work and were then assigned to our tents.

This air field was simple. It was originally a large farm area, with a runway built down the middle. The ground had been leveled by bulldozers, and the run way built of inter- connecting sheets of steel, just as we had seen at the other bases. The runway was shorter and less wide than permanent concrete runways, but this construction design was fast and simple to build. All this allowed for easy change of air base location if necessary. A little less than halfway down the runway and about 30 yards to the side was the control tower. It was a one story wooden building, sitting on pilings to a height of two stories. Around the outside walls was an open porch where the commanders could monitor the mission take off and return. The taxi strips and revetments were built similar to the runway. This was the center piece of Celone Air Base.

On one side of the field were the Americans flying B-17 Flying Fortresses and on the other side were the South Africans flying B-24's. The service groups, including armament, engineering, communications, and ordnance were all loocated close to the runway and airplanes. Each squadron contained these service groups as parts of its organization. In addition each squadron had orderly, supply, transportation, medical, S-2 (intelligence), photo and mail room sections. The four squadrons of the 463rd. Bomb Group were the 772nd., 773rd., 774th., and the 775th. Each squadron occupied its own area of the field. Everyone lived in a tent with the officers in one area and the enlisted men in another. The tents were all placed in a line about 20 feet apart. Each squadron had its own mess hall, enlisted men's club, and officer's club as well as a small parade ground.

We were all assigned to our tents. We gathered our personal bags and off we went to the supply room to pick up canvas folding cots, pillows, and blankets. Moran, Sizer, Manion and I were assigned to the same tent. It had been occupied by the officers of a crew that had been shot down, and so it had not been occupied for awhile. As was usual when this happened, others "permanently borrowed" things that were scarce, such as wood. By this time night had set in, and we were finding our way around by flashlight. When we reached our tent, we were shocked. The tent was leaning, with a muddy dirt floor, no heat and no light bulb. Needless to say, there was much cursing going on. We were tired, hungry, and cold. Besides, we were officers and were here to end the war and thought we should be received better. With our flashlights we found a spot in the tent to place our cots and sleeping bags. We could not even find a dry spot for our bags. We all fell asleep, but, during the night, the wind became strong and it began raining. The weather was cold. With that, the entire tent collapsed on top of us. Since everyone was OK, we decided to prop up the tent just enough to get the weight of the tent off of us. We went back to sleep with plans to fix it properly in the morning.

About 3 AM we were awakened by the chatter of men walking by on their way to the mess hall. At that time the 463rd. was flying out of the 301st. airfield while our runway repair was underway. So the airmen needed to get up earlier than usual since they had to be transported by truck to their airplanes over at the 301st. airbase. As these guys walked by, they commented on the fallen tent, that a new crew had moved in last night, and "the poor bastards had a lot to learn". After things quieted down, we again went back to sleep, with the luxury of sleeping until dawn. Since we had slept in our clothes, partly because there was no dry spot to place them, getting dressed was putting on our boots and coats. After struggling to find a way out of the collapsed tent, we surveyed the damage and decided the best thing to do was to go to the mess hall for breakfast. On our way to the mess hall, we all agreed that, as unpleasant as last night was, it was still better than the infantry who had to sleep in a fox hole while being shot at. At breakfast the veterans advised us about our future combat lifestyle. It is possible they exaggerated a little.

On returning to the tent and surveying the situation a little better, we concluded it needed to be completely taken down and re-built. Although we put up tents when in boot camp, none of us felt very confident about this task. We agreed on our assignments, that Moran would check on the enlisted men to be sure they were OK and that Sizer, Manion, and I would visit officers at other tents to find out what we needed and where to get it. They were all helpful and sympathetic to our situation. In addition to reconstructuring our tent, we needed to fix our heater, to build a wooden floor, and find a water supply and a fuel tank for the heater. Sizer, Manion and I were no sooner back to the tent when Bob Moran arrived with the rest of the crew. They had all volunteered to give us a hand. They were fortunate to be assigned to tents which were functional and had all the necessities. Four of them were in one tent and two were in another. They laughed at the condition of ours. They took over and in no time at all, they had re-erected our tent, and it was a masterpiece in our eyes.

Wood was at a premium as there was none to be found. The enlisted men in the crew reassured us, "Don't worry" they said, "we'll find some." That night in the early morning hours, we hear someone in our tent calling to us. There they stood with arms full of wooden planks. They were used and muddy on one side. "Let's put the floor down", they said. In a few minutes it was all in place but not enough. "We'll have more tomorrow". they said and left. We all went back to bed for a few more hours of sleep. The next day, a notice posted on the bulletin board stated that the 774th. squadron reported wood being stolen the night before and suspected the 775th., and that this would not be tolerated. We had not asked where the wood was from, and we did not want to know, being only too glad to have it. Bob Moran mentioned the notice to the enlisted men. They laughed and could not believe someone would steal wood. We "knew" Dawson, Carrick, Mills, Fouts, Holly, Fotos, and Shirley would not do such a thing. That night, again in the early morning hours, another a load of planks arrived. This completed the job with a little wood to spare. The following day, another notice on the bulletin board appeared, this time stating that wood was stolen from the 773rd. squadron. I asked another pilot who was reading the bulletin board about it and he laughed. "This goes on all the time and is like a game. Every time a crew is shot down or re-assigned their tent is cleaned out." The end result was that we had a nice wooden floor in our tent.

Our tent and home

Home of Mills, Carrick and Dawson

Water and lighting were no problem. One trip to the Supply Room was sufficient to get one 25 watt bulb and two five gallon tanks of water. Each tent was was only permitted to have one 25 watt bulb . This, I learned later, was because of the primitive electrical wiring supplying the tents. A single power line ran along the top of each tent and all the lights were in series. A twenty five watt bulb does not give off much light and reading at night required using a flashlight. I later wrote to Mother asking that she send me some light bulbs. About 6 weeks later a care package from her arrived containing two 150 watt bulbs. We were elated until that night. When the electricity went on, we screwed the 150 watt bulb in and all of the lights down the line from our tent went out. The yelling was terrible. The lights frequently went out, and so no one was suspicious initially. Our tent lit up like the moon, and so our secret was out. There went our 150 watt bulbs.

Three times a week, a water truck refilled the empty water tanks located outside the tent. We used our steel helmet for hand and face washing as well as shaving. We made a wash stand out of wood with a hole cut into it in which the helmet fit. This held the helmet stable while filling it with water. We heated water on the stove using a mess kit cup to hold the water.

Our stove, as all others, was made from half of a 55 gallon drum with a 4x4 inch hole cut out of the open edge. On the closed side of the drum, at its periphery, a 4 inch diameter circular hole was cut with a one inch high rim welded to it. This is where the smoke stack fit which extended thru the roof of the tent. The smoke stack was a 4 inch metal pipe. The fuel tank was an airplane wing tank holding 100 gallons. It was placed outside the tent on a frame which elevated it 2 feet off the ground. A quarter inch copper tubing was attached to the tank and ran 2-3 inches underground into the tent to the stove. There was a stopcock about one foot before the open bottom of the stove to control the flow of fuel. The stove sat on sand, gravel, and small pieces of brick. Once a week, a fuel truck filled the tank with aviation fuel. These stoves really put out the heat. When turned up, the entire drum would become red hot. We also used the stove for cooking food we received in care packages. These stoves were efficient but dangerous particularly when lighting. This was done by lighting the end of a rolled piece of paper or a wooden stick, then putting it into the opening of the stove at the end of the tubing, sort of like lighting a pilot light. When it lit, there was a small "boom". The severity and loudness of the boom was directly related to the amount of fuel that had entered the stove before lighting. Singed eyebrows were common and explosions almost daily. Once our stove lifted 4-6 inches off the ground, and the smoke stack disconnected from the stove due to the explosion when lighting it. However, it sure kept our tent warm and our coffee hot.

The showers were located between the enlisted men's area and that of the officers. Since water was scarce and had to be trucked in, we were allowed only one shower a week, and that was a cold shower. Needless to say, we wasted very little water in cold showering in the middle of winter in a shower room without a roof. We showered in groups that were scheduled ahead of time. With showers that infrequent, even a quick cold shower was a treat. The toilets were outdoor privies with 4 holes, which sat over a slit trench. These were scattered around the squadron. Nobody bothered with the privy for urination. There were no women on our base.

The officer's mess hall was about the equivalent of 2 blocks from our tent. It was open at the required times but also open to serve snacks to crews just back from missions. The officer's mess hall was a limestone building with one large dining room with rows of tables, each of which accommodated 8, and the food was served cafeteria style. The enlisted men's mess hall was similar. The meals were routine. Breakfast was powdered egg scrambled, coffee, powdered milk, toast, butter and hash brown potatoes. Powdered eggs were not a culinary delight. Lunch was usually spam and cheese sandwich, tea or coffee, and sometimes lemonade. The butter was more of a spread with a cheese taste. The food was selected by the Army Air Force to minimize risk of spoilage. The evening meal was spam, stew, or liver. For the first time in my life, I learned to like liver and have enjoyed liver ever since. We had all the sugar and coffee we wanted. Back in the States, these and many other foods were rationed, and everybody needed stamps to buy his or her allotment. We often complained about the food but realized at least we had hot food that was more tasty than the infantry had. Spam was pervasive. In fact the care packages I received from Mother even contained spam. The one thing we had a surplus of and had our fill of was spam. She continued to send it to me until I finally wrote and asked that she not send anymore. With rationing on the home front, it was difficult to send candy and cookies so she was sending what she could get hoping that I would enjoy it.

I remember an incident having to do with the kitchen. One day Sizer, Moran, and I were approaching the mess hall for lunch. As we approached, we heard a loud "whoof", followed by shouting and saw flames coming from the kitchen. Almost simultaneously, a young boy, 12 or 13 years old, came running from the kitchen engulfed in flames. He collapsed about ten feet from the kitchen and the military kitchen personnel who were running behind him yelling for him to stop, immediately put out the flames by wrapping him in their coats. By the time that we were on the scene, his pupils were dilated, he was comatose and it was clear that he had second and third degree burns over his entire body, face, and extremities. He was in shock. An ambulance with medics arrived immediately. He was rushed to the base hospital, but he was dead on arrival.

Apparently, he was a kitchen civilian employee. He was told to clean some cooking grease off the floor. He, on his own, got a can of gasoline to do the job. He opened the can and poured some on the spot. The fumes spread to the stove and immediately lit, and the can exploded dousing him in flaming gasoline. This was my first witness of death, and it was not pleasant. I had seen him around the mess hall daily, and he was such a nice pleasant Italian boy. All the fathers were away serving in the Italian Army, and so the young boys worked when possible in jobs with the American military. This allowed them to bring home some money and other goods such as K rations, gum, cigarettes, etc.

After arrival at 463rd. Air Base, all of the combat equipment we received before departing for overseas was checked again to be sure it was functional. Also they issued us more ammunition for our .45 caliber pistols. Our emergency kits contained fishing line and hooks, a small compass, and a large knife with two 6" blades, one of which was a knife blade and the other was a saw blade. The kit also included a packet of sulfa drug, a bandage, a tourniquet, an ampule of morphine, and a silk map of Italy, Yugoslavia, Austria, and Germany. It also contained a letter written in Italian, German, Russian, and French stating that anybody who aided a US airman would be compensated by the US government. I also added to my kit a toothbrush and toothpaste. Bob Moran kidded me about that the rest of the war and after. He said if ever we were taken prisoner, I would have the cleanest teeth among the POWs. All airmen carried these kits on every mission. During our combat training in the States, we were taught how to avoid capture and survival if we were shot down. By League of Nations agreement, POWs were only required to give name, rank and serial number. The U S military also recommended that captured American soldiers should try to escape if possible.

During indoctrination on combat protocol, we flew some local flights to become familiar with the local area, the landing strip, and also to give the crew some gunnery practice. The routine that had been developed was that the new crew would fly up to the first 5 combat missions with another veteran pilot at the pilot's controls and the regular pilot would fly as co-pilot. The regular co-pilot would fly as co-pilot with another crew. This procedure familiarized new crews with combat so they were less likely to make any disastrous mistakes later when they are on their own.

I remember the events of our first combat mission. Every evening on leaving the mess hall, we checked the bulletin board at the Orderly Room to read the flight schedule and airplane assignment for the next day. The schedule listed the crew by name and which airplane they would fly. The planned target and take off time was not listed, only the airplane number and its crew. Those bomber crews who were assigned to fly the next day, would go back to their tents to check the equipment in their flight bag to be sure it was all there and in good condition. The next couple of hours were free. Some used the time to write letters, some went to the Chapel, some read and others played cribbage or cards. No drinking was allowed the night before a mission, and most crewmen went to bed early. The flight crews were awakened at 4:30 AM. They then dressed in long underwear, flight suit, put on a flight jacket and went for breakfast. Over breakfast some talked about home, some talked about the mission and whether it would be a "milk run" or one of the toughies --Polesti or Vienna or some other challenging target.

The crews then returned to their tents to use the latrine and collect their gear. We wore woolen socks, a jacket or sweater under our fleece lined leather flight jackets and fleece lined pants as well as fleece lined boots. We used scarfs, aviators' helmets, woolen under glove and leather over glove. When we entered enemy territory we all put on a steel helmet over our flight helmet. Some airmen chose to put on their heavy flight gear at the airplane. We all carried our flight bags with parachute, emergency kit, pistol, sandwich or two, and a thermos of coffee or tea which we picked up at breakfast. Carbonated drinks would break the themos at high altitude.

The crews flying a mission that day met at the squadron orderly room where trucks picked us up and drove us to group headquarters for briefing. The crewmen were generally quiet at this time, and the smokers were puffing away. On arrival at group headquarters, we climbed out of the trucks leaving our flight bags behind and went into the large briefing room. After everyone assembled, the group commander walked in and everyone stood up at attention. He immediately gave a "be seated order". He said a few words and told us the day's target and its location. He told us why the target was important, gave us words of encouragement, and wished us luck. Next the pilot of the group's lead airplane spoke. He unveiled a large wall map, pointed out the target, our flight plan to the target, the altitude and compass heading of the bomb run, and the time on the bomb run. He also gave the expected duration of the mission. Next, a meteorologist described the weather conditions we could expect, and finally, an intelligence officer briefed us on the expected severity of flak and enemy fighters and where this was most likely to occur. He pointed out the present location of the eastern and western fronts, what would be our best escape routes if shot down, and the location of underground activity. Following this, a Chaplain led the group in prayer, said a few words and blessed us all. After this, the officers, radio operators and togeliers remained, breaking up into the respective groups for further specific briefing. The pilots were told which position in the squadron formation his airplane would fly in, the position of his squadron in the group formation and the time we should start our engines. We also learned more specific information about the mission. We then went back to the trucks and rode out to our respective airplanes. On arriving at the airplane, everyone put on the rest of their heavy flight gear and parachutes. After a few words as a crew, we all boarded the airplane and began checking the equipment. At the designated time, each squadron started their airplanes' engines. The crew then checked in over the intercom one at a time on request from the pilot. Each crewman reported any problems or that all was well and ready to go.

We then taxied out to the end of the runway, and, with a green flare from the tower, take offs began at 15 second intervals. The squadron lead airplane made a slow climbing left turn, and each succeeding airplane would do the same, flying to the assigned position in the formation. The squadron flew out over the Adriatic Sea at which time the group formation would be complete. All gunners test fired their guns when the formation was over the Adriatic. The formation always assembled and flew north over the Adriatic Sea to avoid enemy flack while gaining altitude. As we neared ten thousand feet of altitude, I instucted the crew over the intercom to go on oxygen. A minute or so later, Bob Moran checked with each crewman to be sure the oxygen system was functioning properly. We flew over the Alps to reach central Europe. We maintained radio silence to avoid giving away our location. The Germans supplied the Italian front using the Bremmer Pass which had a good highway and raiload running through the Alps. They used it in the past to supply Africa via Italian ports. It was well fortified with many anti-aircraft guns. We avoided this area when we could unless it was the bombing target. However, we periodically ran into unexpected anti-aircraft fire on the way to other targets as well. Enemy fighters had become less of a threat at this stage of the war because of Germany's fuel shortage.

As we approached the target area, the lead airplane broke radio silence to notify the others to tighten the formation and prepare for the bomb run. We were flying at 30,000 feet as we had been ever since leaving the Adriatic. Once we started the bomb run, the heading and altitude had to be steadily maintained in order to have a successful mission. This was the most dangerous part of the mission because the anti-aircraft guns were concentrated around the important targets which were the ones to be bombed. The anti-aircraft guns were coordinated with radar which determined our altitude and heading, and the shells were set to explode at that altitude. To try to throw the radar off, we would throw chaff out of the side door to fool the radar operators. Chaff was strips of aluminum foil, which reflects the radar beams and confuses the enemy as to our true altitude. The waist gunners, Tim Holly and Nicholas Fotos, were responsible for this important task and did a good job of it. The flak explosions could be heard and seen as a fire ball and a puff of smoke. The flak explosions near us jolted the airplane, and sometimes we could hear the shrapnel tearing through the airplane. We had to continue the flight formation hoping that no one and no vital part of the airplane was hit. We could not take evasive action since flying straight and level was the only way a mission could be successful. If cloud cover prevented us from seeing the target or the lead bombardier could not get a good site on the target, then we went around and started a second bomb run. This was what everyone dreaded because the second time around allowed the enemy gunners to really zero in on us. Fortunately, this happened only twice on the missions that we flew. We opened the bombay door early in the bomb run. Hearing the bombardier say "bombs away" and seeing the bombs fall from the airplanes of our formation was like music to our ears. Following the "bombs away" the radio operator on our crew, Don Carrick, would visually check the bomb racks to be sure all the bombs had fallen out of the bombbay . He would confirm this to the crew over the intercom. The bomb run was always a tense and anxious time.

Vapor trails from our formation on route to the target

As soon as the bombs cleared the airplanes, we started a turning decent of about 2000 feet to avoid continued flak which had now zeroed in on our altitude. Soon we were out of the flak heading home. The crew members would check in to report injuries or airplane damage. Everyone was still on oxygen at that point. We maintained the altitude to which we had dropped following our bomb run and continued to head home. All crew members were alerted to observe the effectiveness of our mission and the amount of damage inflicted on the enemy target. During the entire flight, the crew members were watchful for the approach of enemy aircraft and the condition of the other airplanes in the formation. As we approached friendly territory, we descended gradually, and the crew went off oxygen as we went below 12,000 feet of altitude. Now the flight became more relaxed, and the crew could move around the airplane more freely. It was now coffee and sandwich time. The coffee was in a thermos and still slightly hot. The sandwiches were kept from freezing by wrapping them in a rags or towels. The temperature was minus 60 degree F. These cold, stale sandwiches sure tasted good at this time and place. Once we were back over the Adriatic, we all felt safe.

As we approached the Celone field, the group broke up into a loose formation and landed according to squadron in sequence. Each squadron flew the downwind leg in formation and at the appropriate moment each airplane would "peel off" from the formation and land. Waiting at the side of the runway were ambulances and fire trucks if needed. After landing, each airplane taxied to its revetment and parked. After gathering our equipment and placing it in our flight bags, we disembarked, gave a verbal report on the airplane's performance and problems to the crew chief then got into a waiting truck. We went to operations for post-mission interrogation by the intelligence officers. As a post-mission routine, each airman was offered 2 ounces of scotch or bourbon whiskey by the Medical Dept. to calm our nerves. If they wanted, the airmen could refuse the drinks then and instead receive credit toward an entire bottle. This is what our crew did. When we had a full bottle, we would have a party. Each crew then rode back to their respective squadron. Most of us dropped off our flight bags at our tents, went to the mess hall for a snack, then took a nap. The crewmen who did not fly the mission that day were usually full of questions concerning the target, how much flack, how many enemy fighters, were any airplanes lost or were there any injuries or deaths. This was basically the routine for all missions. I describe some instances on certain missions later. We were told that in January prior to our arrival that the number of missions required to complete a tour of duty had been decreased from 50 to 35. Needless to say this made the veterans very happy.

The first mission our crew was assigned was to Vienna, Austria. I was assigned to fly as co-pilot with the rest of the crew. Bob Moran was to fly co-pilot with another crew. Our pilot was a combat veteran pilot whose job was to give us on the job combat training. We were all anxious in anticipation of our first combat experience. At briefing we learned that there was possible bad weather over the Alps. Everything went well until we were 90 minutes into the mission, when it was aborted because of severe weather over the Alps. The formation made a 180 degree turn and headed back to Celone Air Base. The fuse pins on the bombs had not been pulled as yet, so the bombs were relatively safe but the fuel tanks were still nearly full. As we approached the field on the downwind leg, the pilot turned to me and said "Now I will show you how a good pilot lands these airplanes". With that, he started a descending 180 degree turn to the runway to land. I thought he started this turn too soon and would not be able to bring the airplane to the end of the runway at landing speed. I suggested we go around, but he thought we could make it. We were half way down the runway and still had not touched down. At this time I said "We can't make it, go around". His pride got the best of his judgment as he proceeded to land. About that time, the wheels touched the runway, and he called for an emergency landing. We were landing on a mud covered, slick runway. The engineers had left a stack of steel matting about 8-10 feet high at the end of the runway. The emergency landing technique involved getting the tail wheel off of the runway to place more weight on the main landing gear and I, as co-pilot, lifting the flaps to place more weight on the front wheels. Simultaneously he was applying full brakes. All this was to no avail. The landing gear just skidded on the runway without noticeably slowing the airplane. It was obvious we were going to hit the stack of steel matting. I called over the intercom to "brace yourselves". Fortunately Roger Sizer and Art Fouts were not in the nose of the plane as they often were on landing. They had previously gone back to the waist section. For a few brief seconds we knew that we were going to crash into the stack of steel matting at the end of the runway and that we could do nothing to prevent the accident except to hope there would be no explosion or fire. The right side of the airplane crashed into the steel matting destroying the number 3 and 4 engines, the propellers, and part of the nose. There were no explosions and no fire, but we immediately evacuated the airplane after we turned off all 4 engines. A fire truck and ambulance were immediately on the scene. There were no injuries but we had an anxious and embarrassed pilot and co-pilot. The crew was told to remain at the site to await an investigating team.

We made sure that there were no fuel leaks and that the bombs were safe. We were carrying 500 pound bombs, and two had detached from the bomb rack on the front end. If the detonator pins had been removed, there would have been a severe explosion from the bombs and the full tanks of fuel. After the remainder of the airplanes landed, the investigating team appeared on the scene. We noticed while waiting that the South Africans in their B-24s began to take off on their mission. We remarked that it must be a long mission or the bomb load was unusually heavy because they used every bit of the runway for take off. A moment later, along came one B-24 which did not leave the runway in the area where the other airplanes did, and it was obvious that it was going to crash into our airplane. All of us who were watching yelled in unison "run" and, with that, took off like Olympic sprinters. As I glanced back, I saw the pilot pull the steering column back into his abdomen, and the airplane shuddered as it left the runway. However, it did not gain sufficient altitude to avoid our plane. The nose wheel of the B-24 hit the right wing of our airplane tearing off the distal third of the wing and tore off the nose gear of the B-24. The airlane lost a few feet of altitude and went skimming along the tall grass and scrub bushes past the end of the runway. I thought for sure the B-24 was going to crash but the pilot did a remarkable job and gradually gained altitude. I learned later that the pilot dropped his bombs in the Adriatic Sea then burned most of his fuel before making a safe planned emergency landing at the 301 airfield. It was a great job of piloting.

We now had an airplane with only 2 engines, part of 1 wing missing and part of the nose gone. The enlisted men were allowed to leave the scene after a brief interview by the investigating team. The pilot and I were on the scene for about 4 hours, and he and I were interviewed separately. It was obvious what had happened: the steel sheets should not have been left at the end of the runway and there was an error in pilot judgement. We were all fortunate that no one was hurt. The pilot and I were told we would be contacted again, but I never was and do not know if the incident resulted in an entry in my record. The pilot was completing his combat rotation, and I doubt he received much of a reprimand since his record had been good. He and I never discussed this incident afterward, and he soon left for the States. I suppose I was exonerated since from then on out I always flew as pilot on our missions. Although this was a trying experience, it was an aborted mission and did not count toward our total.

On February 13, 1945, we learned from the bulletin board that we would be on a mission the next day. We were assigned to fly as an intact crew, making us all happy. There were few missions the previous week because of inclement weather. We learned that our target was the marshaling yards in Vienna. We were not too pleased about this because we had heard from the veterans that the flak over Vienna was heavy and accurate. The last airplane in the 775th. Sq. that had been lost was shot down over Vienna on January 15. As we thought about it, we realized that we may now be sleeping in what was their tent. Anyway, we had no choice. Prior to boarding the airplane, I told the crew, "this is what we have been training for. Remember to do the things we were trained to do and we will be OK". This was an exciting moment with mixed emotions. I will always remember getting in the airplane, the crew checking in, starting the engines, taxing to the end of the runway, then moving into position on the end of the runway waiting for take off. I thought, "I hope I do everything well". We took off at 15 second intervals, and Bob Moran was timing. He gave me the signal and we started our take off roll. We were assigned to fly in the number 7 position in the squadron which was the usual position for a first time combat pilot because it allowed more maneuvering room for the airplane. As the lead airplane reached the downwind leg of the slow climbing turn, all seven airplanes were in a loose formation. As we tightened the formation, the squadron leader brought us into the group formation. As we flew north over the Adriatic Sea, the gunners loaded their guns and fired a few rounds to be sure they were operational.

On my last leave home, my sister Dorothy gave me a simple 127 mm camera which required no focusing, only pointing and clicking. I carried this on every mission and got some excellent pictures. Once in flight, I placed it in easy reach above the instrument panel. Don Carrick used a government issue Air Force camera which he checked out from a friend in Intelligence. Of course he got better pictures, but his selection was more limited than mine since his camera was fixed to the airplane.

We were soon at 31,000 feet of altitude having gone on oxygen at 10,000 feet. The temperature was - 60 degrees F. B-17s are wide open and not pressurized, but our fleece lined flight suits kept us warm but were bulky. As we crossed into enemy territory, we donned our steel helmets over our flight helmets. The remainder of the crew also put on a flak vest. The pilot and co-pilot could not as it interfered with our jobs. The flak vest was similar to a bullet proof vest but the vest was made of flat pieces of steel sewn in the vest like the scales on a fish. The flak vest gave protection against flak shrapnel but were heavy, bulky, and restrictive. The Alps were beautiful, high, rugged, and snow covered. I had never seen mountains so huge before. Roger Sizer, our navigator, kept us informed as to our location, and if shot down, the direction to travel to get to friendly territory. He also pointed out various cities as we flew over them. P-38s and P-51s had escorted us into enemy territory, but they reached their limit on fuel and so had to turn back.

Everything was beautiful, the weather, no enemy fighters and what an easy mission, or so we thought. Things then changed rapidly as we approached Vienna. Flak was everywhere. The flak was initially low since the enemy gunners had not yet figured out our true altitude. The waist gunners, Fotos and Holly, threw out chaff. We then received a radio command to tighten the formation and begin the bomb run. This meant we had to fly straight and level regardless of the flak, which was now at our altitude, with black puffs all around us. We felt and heard the explosions near our airplane. We opened the bombay doors, and we soon could see the bombs fall from the airplanes in front of us in the formation. As we heard "bombs away", we immediately started our descending turn and in a minute or so we were out of the range of the flak. All crew members reported in "all OK". Those crew members who could see the bombs explode took note, so they could later report to intelligence. Don Carrick checked the bombay to be sure that all of the bombs had left the airplane. The crew members' reports indicated that the mission was a success

As we flew closer to home, we were again escorted by P-38 fighters which stayed with us until we reached friendly territory. We spotted no enemy fighters on that mission. Possibly they were engaged by our fighter escorts before reaching us but I would not know. When we descended enough to come off oxygen, we reached for our sandwiches, only to find the sandwiches frozen. We obviously had not realized they needed to be protected from the cold. We landed safely, reported problems with the airplane to the ground crew, and gathered our equipment. After viewing damage to the airplane, we boarded a truck and rode to the debriefing. The mission was declared a success. Our crew had no injuries, and the airplane had only some minor damage from shrapnel. Our first completed combat mission was now over, and I was extremely proud of the performance and bravery of the crew. We were one happy bunch. The flying time of this mission was 6 hours, 55 minutes.

Our combat missions varied in duration and severity, but the morale of our crew remained good, thanks to their character and sense of humor. There were some sad times and some thankful times, but everyone remained focused on the missions. Our flights were not regular and depended on the unpredictable weather. However, the squadron was always in a state of readiness. Our squadron frequently flew 12 airplanes on a mission, while other days only 6 or 7. Up to the time we joined the 463rd. Bomb Group and for awhile after, all bombing was done visually using the Norden Bomb Sight. On every mission a primary and secondary target was assigned in case cloud cover obscured the primary target. In that event, the mission was then shifted to the secondary target. As previously mentioned, our airplane, the B-17G, number 8582, was equipped with radar which allowed sighting the target and bombing it through cloud cover. This equipment was being adopted for combat shortly after our arrival and proved to be very successful.

To allow for take off in the presence of severe fog covering the field, a pipe carrying aviation gasoline ran along each side of the runway. A gasoline fire would be lighted along each side of the runway which would "burn" off the fog over the runway and allow the airplanes to take off and fly the mission.

Togglier Art Fouts combat ready

Another problem was the flight clothing. At 30,000 to 32,000 feet of altitudes on missions lasting from 6 to 10 hours, the temperatures ranged from minus 10 to minus 60 degrees depending on the time of the year and the ground temperature. The bulkiness of the flight suit created a real problem but was the best available at the time. Maneuvering in the confined space of a B-17 with all these bulky clothes on was not easy. Trying to relieve one's self of urine in a relief tube in the tight space of the airplane at minus 40 degree F temperature on an 8 hour mission is not easy. Ask any WW II aviator. Another problem was that the 50 caliber machine guns frequently jammed. The gunner was not able to unjam the gun with his bulky gloves on, and so he would have to take off his gloves. Often when he touched the cold metal of the gun with his bare hand, his skin would become frozen to the gun metal. His choices were limited. If he attempted to pull his hand away from the gun, often the skin remained on the gun. If he could reach his thermos, he could poor hot coffee on his hand to loosen the skin, or he could urinate on his hand in an effort to get his hand free. This was a real problem and finally the Air Force began to issue a silk, thin, tight-fitting glove which airmen wore under the thick woolen glove and the leather outer glove. This meant wearing three pairs of gloves but it solved the problem. Toward the end of the war, the Air Force issued electric flying suits which included electric gloves and boots. The electric clothes were like an electric blanket with thin electric wires running through out it. The airmen would just plug it in once the engines started and set the thermostat at the desired temperature. These were great, warm, and light weight. But If the airplane lost electrical power, they were not very warm.

Our next mission was to Landshut, Germany, an 8 hour 15 minute flight in which we again had much flak. An incident happened on this mission that I later learned about. On our way to the target, Al Dawson, the ball turret gunner, told Bob Moran that he would be off the intercom for a few minutes while he used the relief tube. On his return to the turret , he stowed the walk around oxygen bottle in its proper place but he found out later that he had not secured it properly. Later in the middle of the bomb run with flak bursting all over the place, the oxygen bottle came loose from the bracket and fell on the ball turret track. As Al swepted the turret around, it tore off the neck of the oxygen bottle. With that, the bottle began flying around the waist section of the airplane, spewing oxygen, stirring up dust, dirt, and candy wrappers as well as clanging against parts of the airplane. The waist gunners, Tim Holley and Nick Fotos, thought we had a direct hit and began yelling so. In a few seconds they realized it was just Dawson's oxygen bottle. Dawson then thought his oxygen line had been severed and was now without oxygen and had only a few minutes to live. I am sure he would have had less than a few minutes to live if Nick and Tim could have gotten to him. He certainly would have received the crew "idiot of the month award" if I had known about it then.

My crew became a highly cohesive unit as we became combat hardened. The enlisted men ate together, lived together, and worked together, as did the officers. As a group we flew together, told stories together, and helped each other. We developed a keen sense of mutual respect especially on the job.

On our return from one mission over Linz, Germany, in which the flak had been very heavy, Don Carrick counted 67 shrapnel holes in the skin of the big bird. With so much shrapnel entering the airplane, I could not help but wonder how we could all escape injury and how the airplane could continue flying. I guess that is why I never counted holes in the skin of the airplane. It's a testament to the engineering marvel of the B-17. At times like this, I would think of my experiences duck hunting when 2 hunters would fire 6 shells at a passing duck, and the duck would continue flying unhurt.

There had to be some divine intervention as we had one experience that I will always remember. Our squadron had one airplane which was thought of as a "bad luck" airplane. Al Dawson remembers it as the "Dirty Devil". This airplane had flown many missions, had been shot up a lot, and often had mechanical problems. It was old but still around. The veteran pilots all said, "Don't fly that airplane because something bad will sure as hell happen to you and your crew". Everyone avoided it. Well, one evening on checking the next day's flight schedule, there we were scheduled to fly the "Dirty Devil". I immediately went to see our Operations Officer and requested another airplane. He informed me that it was all we had ready to fly the next day. I got his OK to switch airplanes if we could find a substitute by morning. I then checked with our Engineer, Archie Mills, and explained the problem. He said he would speak to one of his friends in maintenance. Perhaps he could get his crew chief friend to put more men on one airplane and have it ready by flight time the next morning. Sure enough, by flight time our airplane assignment had been changed and we flew the mission without difficulty. That morning before take off, another crew had a problem with their assigned airplane and had to be shifted over to the "Dirty Devil" since it was the only one available. It took off after us and I learned later that day after returning from the mission that it had crashed on take off and the entire crew was killed. We were scheduled to fly that airplane and did not. These things get one thinking, and it is understandable why there are so many superstitions among the airmen. Most aviators just liked to fly the same airplane if possible.

On one 8 hour mission to Rhuland, Germany, we had the usual flak but did not think we had airplane damage. We were flying in the number 3 position in Able Squadron. Everything had gone well for us but in preparing to land we found our landing gear indicator was nonfunctional. We were not sure if our landing gear were down or not. I called the tower, told them the problem, and that I would do a "fly by" so they could check. The tower responded "Roger, Able 3. Do a fly by at 200 feet." We then flew over the runway at 200 feet of altitude. On completing the fly by, the tower told us that the right landing gear was not down at all. They told us to leave the landing pattern, climb to 5000 feet, and try to manually lower the gear while the rest of the group is brought in. Dawson had to go into his ball turret to observe the landing gear while Archie Mills and Don Carrick cranked it down. The crank was stored on the forward bulkhead of the bombbay area. Standing on the bombbay cat walk, they succeeded in cranking the right landing gear down. Dawson reported that both landing gear were down and looked OK to him. Moran then notified the tower of our success and asked for instructions. They told us to do another fly by. After the fly by, the tower gave us permission to land. The crash truck and ambulance were standing by. Moran instructed the crew to assume crash landing positions. We landed without difficulty, and the landing gear held up fine. We later found out that shrapnel had knocked out the electric motor to the right landing gear.

Our navigator, Roger Sizer, was assigned to another crew for one mission because their navigator was grounded due to an illness, and my crew was not scheduled to fly that day. Roger got up early in the morning while the rest of us slept on. About 4 hours later, Roger came rushing into the tent obviously very upset and threw his flight bag on his cot saying in a loud voice "I will never do this again. They can court marshal me but I'm not flying with another crew. We were almost killed." After we calmed him down, he related the incident. They were assigned to fly the number 7 position which was the lowest in the squadron formation. On take off, he was in the nose of the airplane working with the navigational charts. After leaving the runway, the pilot was maneuvering to get into formation. However the formation was not gaining altitude. As a result, the airplane Roger was in was losing altitude to the extent that they were skimming the ground, hitting small trees and tall bushes. Branches became lodged between the nose guns and fuselage. Roger was just sitting there in the nose watching all this. Roger said all he could think about were his two children while this was going on. The pilot was able to gain some altitude and break from the formation. He reported this to the lead airplane and was ordered back to the base. The rest of the squadron continued on its mission. I am sure Roger has never forgotten this frightening incident.

When we were not flying, we played volleyball, touch football, or pitched horse shoes. The indoors activities were reading, cards, or cribbage. I took as many pictures as my film supply allowed, which was not a lot. There was no corner drugstore, and so my supply of film was only what I brought from stateside, what Mother could send, and availability at a PX. We had no Fox Photo with 1 hour developing, so I had to develop and print the pictures myself. We had a squadron photo lab which I could use, and the personnel working there taught me lots. I took some good pictures, even though only black and white was available at the time.

We adopted a French Poodle which was a real companion and much fun. We named him Puddles, and he loved to be cuddled. We brought food from the mess hall for him. On returning from a mission, Puddles would always be lying at the tent entrance waiting for us and would express his happiness for our safe return by jumping all over us.

Roger Sizer with Puddles

We employed a 12 year old Italian boy as our tent boy. His name was Willie. He also worked for a few guys in a few other tents. He ran errands, kept the tent clean, shined our shoes, and took our clothes home for his mother to wash and iron. He was a nice kid and did speak broken english. We paid him in Italian Lire occupation money, with some extra pay for specific duties like the laundry. We also shared care packages, cigarettes, and candy with him and his family. Some years after the war, some members of our bomb group traveled to Foggia to see the old air field. Willie was one of their guides. Willie had become a Colonel in the Italian Air Force and was married with children. He enjoyed the visit and the memories as much as the group members did.

Tent boy Willie in front of our tent

Foggia was a city of about 100,000 people, located about 3 to 5 miles from our base. It was mostly in ruins having sustained the ravages of war. There was rubble everywhere and very few completely intact buildings. All bars and clubs were off limits to the Americans, and the MPs had posted "OFF LIMITS" signs. We were allowed to go in shops and barber shops. There was nothing of interest in Foggia, but the people were nice and certainly did not show any outward resentment toward the Americans despite the fact that we were still fighting them in northern Italy. It was not too much later that the Italian Partisans captured and executed Mussolini. Shortly afterward, Italy surrended but the Germans continued to fight on the Italian front. One could go into a barber shop and get a haircut, hair wash, shampoo, shave, manicure, and scalp massage for $10 plus a $ 2 tip. Now this was a treat. Any time I went into Foggia, this was my priority.

About midway through our stay at Celone Air Field, Bob Moran arranged for us to move into a limestone building. How he did it I do not know. It was originally built for the squadron commander. I think the new CO did not want to leave the other officers of his crew and so passed it up since it was too small to accommodate four. It was smaller than our tent but much nicer and with 2 small windows. Roger, Bob, and I moved in since Manion was previously reassigned. It was on the parade ground and close to the mess hall and orderly room. We had accumulated our ration of liquor, and so we decided to have a party. We invited a group of South African officers and some friends. The South Africans brought brandy which was their usual drink. The party proceeded to get loud and high. Bob Moran finally went to bed but periodically complained of the noise. The next time he fell asleep, 4 of us picked up his cot while he was sleeping on it and moved it out in the parade area, covered him with extra blankets and left him there. It was cold at that time. The next morning he was still sleeping outside completely covered, head and all. We could see the other officers going by on their way to the mess hall. Their comments about who it was and why he was out there were hilarious. Finally he woke up, lifted his head, and looked around. He waited until no one was around, got up dressed only in undershorts, picked up his cot and blankets, and came back into our cottage cursing and mad as hell. Bob accused us of putting him out there. Imagine that! We told him that he had become angry the night before because we were making too much noise. So he picked up his cot and moved it outside. He believed this until a few months later when we told him the truth. He was angrier still.

After 10 missions, we went on R & R (rest and relaxation) to Naples. Our group had a crew fly us to Naples, and we were put up in a very nice hotel. Sizer, Moran, and

I shared a room with a large bathroom and a very large bathtub complete with hot and cold water. The first thing was for each of us to take a one hour bath with the tub filled to the top while we sipped vermouth on the rocks and listened to American music. This was so enjoyable that it became a daily ritual. This hotel was American controlled and operated only for American junior officers on R & R. There was a nightly dance from 8 to 12 PM with live music. The girls were all invited by the hotel management and at the conclusion of the dance had to leave the hotel unescorted. Of course one could always arrange a date for the next day or else where after the dance. The dances were fun with good music and refreshments. The tours of Naples were nice and a day on the Isle of Capri was great. What a beautiful place. The water was crystal clear to a depth of 40 feet.

We had brought along candies and cigarettes for trading. One day Bob and I were walking down the street when a young boy began pulling at my sleeve saying, "Hey Joe, trade for cigarettes." He was holding up what looked like a man's gold ring with a huge diamond. I ignored him initially but he continued. I thought, this could be real considering all the thievery going on. We stepped into an alley, and I swapped a carton of cigarettes for this gem. I put it on my finger. It was a perfect fit, and Bob and I walked on. We walked around town sightseeing on this beautiful sunny day. After a few hours, we decided to step into a bar for a cold beer. While drinking the beer, Bob began to laugh and said "look at your finger". I looked at my sweaty finger and it was completely green around the ring. I argued with Bob that the diamond was real and only the gold was fake. He said "OK lets put it to the test and etch some glass." After a one dollar bet about its authenticity, the test revealed that the "diamond" shattered and the glass remained intact. I then realized that the black market was illegal not only to protect the civilians but the serviceman, too . I often wondered what else I could have had for that carton of cigarettes. We saw an opera in a beautiful opera house and toured many museums and churches. The beautiful buildings were spared from the ravages of war. While in Naples, I ate my first pizza ever. The 4 day R & R was great. We did not get much rest but lots of relaxation. We flew back to Foggia to resume duty.

Early in our combat tour of duty, the Germans were still sending up fighter planes to fight off bombing raids, and at the time the Allies were still bombing their air fields. As the war went on, the German fuel supply dwindled to the point that the Luftwaffe could no longer fly. Toward the end of the war we would fly by German air fields loaded with fighter planes and none would leave the ground to fight for the lack of fuel. I remember the first jet airplane I ever saw. We were on a mission to Austria and one came up to attack the group but was intercepted by our fighter escort. It was very much faster than our fighters, and it shot one down before breaking off. The Germans had a few at the time and was just combat testing them. It was too little, too late for their cause.

One afternoon my old friend, Bigelow, from cadet training days, showed up for an unexpected visit. He was flying P-38s and was stationed in the Foggia area. He had escorted some of our missions. We had a great visit, and brought each other up to date. From then on we visited each other periodically even to the point of his buzzing our tent in his P-38 and I buzzing his tent in a B-17. Of course the B-17 was capable of blowing over his tent which neither he nor his commanding officer liked very much. On one visit I was lucky enough to see up close the newly developed American jet fighter which had just arrived on their base. It was slick and not any larger than the P-51. It was there for field testing but not for actual combat duty.

Intelligence told us that the Germans had developed an anti-aircraft missile and were beginning to use them in combat. I witnessed its use on one occasion and did not like what I saw. This was on a mission over Germany and off in the distance at the same altitude as us at 9 o'clock, I happened to see a missile rising to another bomb group. It seemed to level off within the group of airplanes and explode. This one missle knocked down three bombers. It was unbelievable and scary. Fortunately the war ended before they were in widespread use.

On a mission over southern Europe, our preflight briefing indicated that the bad weather over southern Europe was moving out and the weather should be good by the time we reached the area. It did not quite work out as predicted. We were clearing the Alps at 28000 feet and flew into a cloud cover. Formation flying became more difficult as the clouds thickened. In order to see the airplane on whose wing we were flying, we progressively had to tighten the formation until we were almost flying wing tip to wing tip. There were plenty of airplanes around but I could only see the wing of the one right next to me. Finally I could not even see that one, and so we had to break off from formation and fly on instruments. Every airplane was having the same problem. Soon the pilot in the command airplane instructed those of us who broke formation to continue on our stated heading, rate of climb, air speed, and to gather in formation once above the clouds. These were tense moments because of the many airplanes in the immediate vicinity and having to rely soley on our instruments. When we broke out of the clouds at 30,000 feet, airplanes were scattered all over, with others popping up out of the clouds. I could identify our group airplanes by our group insignia painted on the tail. We then rejoined with other airplanes of the 463rd. Group. The airplanes were now mixed up with airplanes from other squadrons as we had to join in formation with whatever airplanes were close by. We had to form order out of chaos. We eventually were all back in formation and continued on our mission, which was a success. That was a hair raising experience and fortunately without accident. By the time of our return, the weather system had moved out, and the flight home was uneventful. We untangled our mixed squadrons after landing. This was an example of the quality of the pilots and co-pilots in our group and the result of excellent training. This incident reminded me so much of the incident I had at Drew Field in Tampa, Florida, that was previously described, in which our good training paid off.

B-17 and flak

Occasionally to add a little spice to the flight, we would tighten our formation position so our wing tip touched the wing tip of the adjacent airplane.

One of the duties of the officers was to censor the out-going mail of the enlisted men. They had been briefed on what could not be included in letters and that the envelopes were not to be sealed until after having been censored. This was sort of a dull assignment but at times funny and other times sad. Servicemen have a real sense of humor. One time the envelope contained two slices of gum. Along with the letter was a note stating that one slice was for the censorer and the other for the person it was addressed to. Others were responses to a "Dear John Letter" or to the death of a close relative. Any information of military significance was cut out of the letter. Censoring was supervised by Lt. Cassidy, our intelligence officer. If we saw any pattern in it then that person was called in and interrogated by Lt. Cassidy.

An exploding target

On one mission to Austria, following the bomb run, when Carrick checked the bombay, he found one bomb was still hung up inside the bombay. This was a live bomb because the fuse pin had already been pulled. After settling down after the bomb run and now over the open country side, Don kicked it out and watched it fall. This was at 30,000 feet and over open country with only an occasional house. Unfortunately that bomb landed on a house blowing it to bits. The intent was to avoid all structures, and the amazing thing was that we could not have hit that house had we tried. We all felt badly and hoped the house was empty.

We again went on R & R after 15 missions and this time to Rome. What a wonderful historic city! Essentially the accomodations were like in Naples but Rome has so much more to see. We took many tours, each better than the previous. It was amazing to see the ancient structures and wonder how they could have been constructed at the time, including the Coloseum, the Aqueducts, and so much more. One wondered what the engineers of that time could have done with present day equipment. The content of the museums was amazing, such beautiful art. The art work in the Vatican is just unbelievable. I was fortunate to be randomly selected as one of a group of about 25 servicemen to have an audience with Pope Pius. I had bought some prayer beads which he blessed. These I kept through the years until recently when I gave them to my daughter Gayle who had converted to Catholicism. We attended an opera in a magnificent opera house. We had box seats with individual stuffed chairs. This R & R ended much too soon. Again the fighting had spared Rome, and the troops bypassed the city.

We were now back at Celone Air Base and back to war. We were getting daily reports on the progress of the ground fighting and obviously things were going well for the Allies. We were now often flying missions in support of the 5th. Army, hitting supply depots, ammunition dumps, communication centers, and transportation centers, all with much success. On May 8th, 1945, I had gone to see a movie on base. The movie had just begun when it was interrupted with an announcement that the Germans had just surrendered and the war was over. The place broke out in bedlam with everyone shouting, hugging each other and running out of the theater. Airmen were shooting their weapons in the air and setting off explosives from the ordnance section. This went on through the rest of the day and night. The very next day some conversation began to shift to what we would each do after leaving the service. Discipline became a little more strict as now everyone had a lot of free time and discipline could quickly deteriorate. The commanders began to develop educational activities to keep the men busy while closing down operations and arranging orderly transportation back to the States.

Our crew had flown 19 completed combat missions all without any injuries. In the process, we established some lifelong friendships. This is common when a group works closely together and their lives depend upon each other. We had many happy experiences considering the circumstances.

The Swoose Group (463rd. Bomb Group) served well and received two Distinguished Unit Citations and for 6 months had the highest bombing accuracy rate in the 15th. Air Force.

All enlisted men of our crew were promoted to Sergeant. Bob Moran was promoted to 2nd. Lieutenant and Roger Sizer and I were promoted to 1st. Lieutenant. All members of the crew had been awarded the Air Medal with one Oak Leaf Cluster.

MISSIONS FLOWN

Date	Target	Flying time
2-14-45	Vienna, Austria	6 hr 55 min.
2-22-45	Landshut, Germany	8 hr 15 min.
2-28-45	Verona, Italy	5 hr 55 min.
3-4-45	Zagreb, Yugoslavia	7 hr 30 min.
3-12-45	Floridsdorf, Austria	6 hr 25 min.
3-16-45	Schwechat, Austria	6 hr 20 min.
3-20-45	Amstetten, Austria	7 hr 10 min.
3-22-45	Rhuland, Germany	8 hr 00 min.
3-25-45	Prague, Czechoslovakia	8 hr 00 min.
3-31-45	Linz, Germany	7 hr 10 min.
4-1-45	Maribor, Yugoslavia	5 hr 40 min.
4-6-45	Verona, Italy	7 hr 00 min.
4-9-45	Northern Italy support of the 8th. Army	5 hr 12 min.
4-10-45	Northern Italy support of the 8th. Army	5 hr 25 min.
4-15-45	Bologna, Italy support of the 5th Army	6 hr 30 min.
4-17-45	Bologna, Italy support of the 5th Army	6 hr 20 min.
4-19-45	Rattenburg, Austria	6 hr 00 min.
4-21-45	Rusenheim, Germany	6 hr 50 min.
4-24-45	Malborghetto, Italy	6 hr 15 min.

In addition, we had 20 local flights which included aborted missions due to weather and training flights. We had the following non-combat flights:

Naples	2
Rome	1
Sicily	1
Piza	1
Milan	1

We knew we had inflicted much damage to the enemy, but we could never be certain of the extent first hand. We saw much damage in the cities but could not be sure what was due to bombing and what was due to artillery. It was not until 1949 that I learned of the enormity of the damage inflicted by the Air Force. That summer I worked as an orderly on a military transport ship carrying refugees from Germany to the United States. Our port in Germany was Bremerhaven and from there I took trips to Bremen and Hamburg. According to the locals, there was no ground fighting in this area, and all the damage was due to bombing. The wharfs were destroyed as were all the nearby buildings. In Hamburg I could stand and look a half mile and not see a single standing structure, only debris and twisted steel. All of the children had been evacuated to various camps in the country, only to return at age 16. On one occasion in Hamburg, the civilians were in bomb shelters continuously for 7 days and nights. The bombing was constant by the American 8th Air Force in the day and the British at night. The Germans I spoke to told me that the around the clock bombing was terrifying.

Chapter 14
Air Transport Command (Post VE Day)

In June of 1945, five members of our crew said goodbye to the other five and to the rest of the 775th. Sq. (Allyn's Irish Orphans) and transferred to Naples. Co-pilot Bob Moran, Navigator Roger Sizer, Pilot Robert Burch, Radio Operator Don Carrick and Engineer Archie Mills transferred. Ball turret gunner Al Dawson, Togelier Art Fouts, Tail gunner George Shirley and Waist gunners Nick Fotos and Tim Holly were transferring back to the States to train for combat in the Pacific. This separation was a sad day for all of us, but we knew that it was coming. Our assignment was to fly 5th Army troops from Naples to Casablanca where they would then return to the States mostly by Pan American Airlines or by the regular Air Transport Command.

We flew B-17s which had been stripped of all armament, their bombays were sealed, as were the side gunnery windows, the ball turrets had been removed, and the radio operator's area was relocated. Benches were placed running along each side of the waist. The bombay was used for baggage storage. We were housed in apartment buildings near the air field, and buses ran from the apartment complex to the mess hall and the flight operations building. The apartments were nice 2 bed room units with a bathroom, kitchen, living room, dining room, and balcony. We became friendly with a very nice Italian family next door. We gave them cigarettes, coffee, sugar, and candy. Once a week, the mother cooked spaghetti and tomato sauce which she shared with us. We often sat on the balcony, had coffee, and spoke about our respective countries. They spoke English well. It was amazing that we were enemies not many months before and now have these friendly conversations.

Our first flight as part of the Air Transport Command was on July 7, 1945, nonstop to Casablanca, a 6 hr. 55 min. flight. We left Naples in the morning, initially flying southwest over the Terrhenian Sea in sight of Sardinia to the west then later Sicily to the east. On reaching the Mediterranean Sea, we changed to a more westerly course over the Straits of Gibraltar with Spain and Gibraltar to the north and Spanish Morocco to the south. Once reaching the Atlantic, we turned south to Casablanca. We unloaded the troops, then flew to Rabat, spent the night, and the next day flew back to Naples. On occasion we flew to Port Lyautey rather than Casablanca and there were times when we spent the night in Casablanca or Port Lyautey. Our flying altitude varied between 8,000 and 10,000 feet. The geography was exciting from the air.

After landing in French Morocco, we had some free time for sight seeing. It was interesting and much like it is portrayed in movies. We saw from the outside spectacular palaces of the Sheiks. I was able to brush up on my high school French while there and did reasonably well. One day the wind was blowing off the Sahara Desert which was a most unusual experience for us. The temperature was about 120 degrees, and it was impossible to ride in an open vehicle without goggles and a scarf to cover the face. I then understood why the Arabs cover their heads and faces while riding camels.

When spending the night in Rabat, the revetments for parking the airplanes were on the far side of the field. The area was secluded, and surrounded by a small river. At night, Arabs would cross the river and burglarize the airplanes. As a result Archie Mills and Don Carrick had to sleep in the airplane. Some time prior, Don had met a 5th. Army soldier who was in a K-9 unit who had a beautiful German Shepherd dog which he was attached to and wanted to take home. We had flown this soldier and others to Naples where they would be boarding a ship for the States. The dog and Don hit it off well and Don offered to buy the dog, but his owner refused. The next day Don happened to run into the soldier with his dog on a street in Naples. He asked Don if he still wanted the dog, since he would not be allowed to take it on board the ship. The owner told Don he wanted to find the dog a good home where he would be well cared for. Don immediately jumped at the chance. The owner refused to take money for the dog and instead instructed Don on how to handle this highly trained animal. Don named the dog Snappie. Snappie did not like civilians and was smart and well trained. He followed Don's commands perfectly. It was amazing how the two blended. Snappie slept on a blanket at the foot of Don's bed and woke up Archie and Don in the morning. He would watch from the balcony for the bus that took everyone to the mess hall, and, when he saw it or heard it, he would bark and go to the door. The three would board the bus and ride to the mess hall. Snappie would be fed at a side door by the mess sergeant, then lay at the main entrance waiting for Don. Even when we were away flying, Snappie would watch for the bus, go to the bus stop, board it, ride to the mess hall, and have his meal, and return on the bus. This was some smart dog.

One day Don came to me with a suggestion. He reminded me of the burglaries of the airplanes that go on in Rabat and suggested we take Snappie along as a watch dog. Don didn't mention this, but he and Archie were getting tired of sleeping in the airplane. I agreed to the idea. After that, Snappie became a fellow aviator. He always laid down next to Don's radio table and did not move unless Don told him to. From then on, Snappie slept in the airplane while Don and Archie were now able to sleep in a bed. No one robbed the airplane while Snappie was on duty. In the mornings, Snappie allowed none of us to board the airplane until Don gave the command.

Snappie watching for the chow bus

One night in Naples, Roger Sizer, Bob Moran, and I went out for dinner and then to a night club. The custom was to serve drinks straight, while the mixers were placed on the table. Back then, soda water came in a quart sized bottle with a CO2 cartridge attached to it with a squirt top on the bottle. It could squirt the water 5 or 6 feet. We were having fun, and Bob and I were kidding around. I squirted him with soda water, and he squirted me. In the process we got some water on the people at the next table, and then they began squirting each other. This progressed until everyone in the place was doing the same. After a few minutes, Roger said, "We better get the hell out of here before the MPs arrive." We were no sooner on the sidewalk when the MPs arrived. We did not stay around to see the end result.

Another evening, Bob and I decided to have dinner at a fancy Naples restaurant located on top of a huge hill overlooking the Bay of Naples. Bob decided to go down to the motor pool to try to get a jeep for that night. He returned saying "everything is arranged and we pick it up in front of the motor pool". Bob was not an experienced driver, and I could not drive at all. We picked up the jeep and drove to the restaurant. The meal was delicious, and the view of Naples and the bay was magnificent. A few hours later as we were driving down the hill, 2 MPs in a jeep pulled up behind us. One was yelling in a megaphone for us to pull over. We were not sure why. Maybe it was Bob's driving or the fact that we were not really authorized to have the jeep. Bob and the sergeant in the motor pool agreed that a jeep would be out front with a key in it, and, if we took it for a few hours and returned it, no one would know. We quickly debated as to whether we should stop or play dumb. We were about half way up another hill when we decided to stop. Bob stayed in the jeep with the wheels turned to the curb and I got out. The MPs were obviously angry by this time, and they were inexperienced besides. They pulled up behind us, both jumped out with their clubs in hand, but they did not apply the emergency brakes, put the jeep in gear, or turn the wheels to the curb. As they were approaching us, I saw their jeep rolling back down the hill. I shouted, "Your jeep is rolling down the hill". They initially ignored me, probably thinking this was a trick. I yelled again, and this time they looked back to see for themselves. They immediately turned to run after their jeep forgetting about us. When the jeep reached the corner, it veered to the side and ran right through the front of a small shop. Bob shouted to me, "Get in, let's get the hell out of here". I hopped in, and we took off. We returned the jeep, went back to our apartment, and we never heard any more about the incident. Some one else must have been on our side.

One day I ran into Glenn Draper, my friend from flight school and B-17 training. He was assigned to 15th. Air Force Headquarters and was the pilot of General Twining's airplane. General Twining was the Commanding General of the 15th. Air Force. Consequently, Draper had a very interesting assignment. Glenn was supposed to fly home by way of Casablanca. But, the night before his departure, he was robbed of everything he had, money, clothes, souvenirs, and even his watch. The only thing he had left was the underwear he slept in that night. To make his flight, he did not have time to buy a new wardrobe, and so I gave him a set of clothes to wear and some toilet articles to tide him over. He promised that the next time he saw me, he would buy me a steak dinner. About 10 years later, he phoned me. He was in New Orleans with his wife and he took Peggy and I to the Monteleone Hotel for that steak dinner. We reminisced and had a great time. That was the last time that I saw Glenn. He passed away about 8 years later.

On August 5, 1945, while flying back from Rabat, I heard over the radio that the atomic bomb was dropped on Hiroshima and that the city was completely destroyed. I could not believe that any weapon that powerful existed. When I arrived in Naples, the Hiroshima bomb was the sole topic of conversation everywhere. No one knew anything about it, what it was, or how it worked. Four days later, August 9, 1945, the second atomic bomb was dropped on Nagasaki. On August 14, 1945, we received word of the Japanese surrender. Everyone was jubilant because we had thought that, after this temporary assignment, we would be going back to the States, to train in B29s to prepare for the Pacific war. The surrender meant going home. We continued flying troops to Casablanca, our last trip being on September 6,1945. On return to Naples, we received orders to prepare to return to the States. The remainder of my crew was going to fly home.

Chapter 15
Flying Home

We were elated to be going home and particularly to be flying ourselves back home. One of our concerns was Snappie. Don Carrick was set on taking him home and the rest of the crew was agreeable. After much discussion and planning, we decided the best approach was not to ask permission but just to take him with us. Don assured us that he could handle Snappie on the flight. The original orders dated September 23, 1945, Special Orders no. 108 had me flying as pilot with a different crew, none of whom I had ever met. I thought this was in error. So I brought it to the attention of someone in headquarters, who corrected the error. Moran, Sizer, Mills, Carrick and I were assigned to fly airplane number 44-8395 a B-17G and will have 10 passengers. On September 24, 1945, we left Naples for Marrakech, French Morocco. This was a 4 hr 15 min. flight on a beautiful day. We landed at the same airfield we landed on during the way over in December, 1944. The next day we had a 4 hr 15 min. flight to Dakar, French West Africa.

In Dakar we billeted in a French Foreign Legion facility. The walls of the facility were about 2 feet thick with huge windows. From the window, we could see the native village outside the grounds of the facility. The houses were round, thatched grass huts built on dirt streets. The village did not seem to have electricity, sewer system, or running water, all very primitive. The next day we took off for our next scheduled stop, Natal, Brazil. One hour out over the Atlantic, I noticed an oil leak in the number 1 engine and a slow drop in oil pressure. With that, Moran and I decided to return to Dakar and had Don notify the Dakar control tower of our decision. We were able to return and land without having to shut down that engine. After landing, Archie Mills, our engineer, immediately went to work on the problem. By that night, he had fixed it and everything was A OK. The next day we again took off for Natal. The flight from Dakar to Natal was over the narrowest part of the Atlantic but still stretched the fuel consumption of the B-17. We were warned of this during the briefing. So we were particularly conscious of our engine settings and air speed. Another thing we learned in Dakar was that the Brazilian government did not allow animals, fruit, or vegetables to be brought into their country. As yet no one knew that Snappie was part of our crew. I brought this to Don's attention but he had already figured out that, if it became necessary to hide Snappie, he would hide him in the space where the ball turret used to be. When the ball turret was removed, the hole in the skin of the airplane was covered over with a sheet of aluminum and the hole in the floor with a piece of wood which was screwed down. Don now made sure that Archie had in his tool box a screw driver that fit those screws. We ran into much bad weather over the Atlantic, and we flew part of the time on instruments.

Our flight took us across the equator into the southern hemisphere. We now became members of a distinct club whose members had crossed the equator. Roger used celestial navigation on this flight. As we approached the last segment of the flight, we began to check our fuel gauges frequently as going through the bad weather caused the airplane to consume more fuel than usual. As the gauges were approaching empty, we spotted land and Natal. Roger, our navigator, had brought us right in on the money. On landing and reaching our revetment, our fuel gauges were on empty. One of the men servicing our airplane told Archie our tanks were essentially empty. He also said one crew was not so lucky and had to ditch in the Atlantic. Fortunately they were rescued.

After we landed but still taxiing, Don put Snappie in the ball turret well with a blanket to lay on and gave him a command to stay and be quiet. He then put the floor covering over him, screwed it down, and put a duffle bag over it. Moran and I had no sooner cut the engines, and the inspectors were on board. They asked if we had fruit, vegetables, or animals on board. We said no, only the personnel and the baggage. They walked through the entire airplane, filled out a form, gave me a copy and left. Snappie did not make a sound. When the coast was clear, Don opened the ball turret compartment and out came Snappie but only on command.

The next day, September 28, we left for Georgetown, British Guinea. It was a four and a half hour flight. Much of the flight was over the Amazon River delta. It was huge in size as is all of Brazil. We spent the night in Georgetown and the next day flew on to Puerto Rico. This was a 2 hr 40 min. flight from Georgetown. We landed in Borinquin which was nice, with good weather. The water around the island was colorful and beautiful.

After a 2 hr and 25 min flight from Puerto Rico, we were in the United States landing at Morrison Field, West Palm Beach, Florida. There we were processed and shipped out. This was the last I saw of my crew while still in the military. Don Carrick shipped Snappie by rail to his home. He placed one of his uniforms in Snappie's cage so that he would have a familiar scent. Don called home to be sure someone in uniform met Snappie at the station to take him to Don's home. After telling everyone goodbye, I left the next day (October 1, 1945) in charge of 4 officers and 70 enlisted men on a train trip to Camp Blanding, Florida. From Camp Blanding, I was transferred to Camp Shelby, Mississippi. On October 4, I received orders to transfer to Kelly Field in San Antonio, Texas, and to report there at the conclusion of a 45 day leave which was to begin immediately.

This was a great 45 days. I visited family and friends, ate well, and caught up on my sleep. Since the end of the war, I had spent much time thinking about my future career. I finally decided that I would try to go to medical school and, if that didn't work out, then I would pursue a degree in aeronautical engineering, for a career in aviation. I consulted with Dr. C C Bass, Dean of Tulane School of Medicine, about my possibilities of getting into Medical School. His advice was to enroll in Tulane to complete premed, make good grades, and give up football. My chances of admission would be reasonable. Medical school then became my objective.

Pat Smith, with whom I corresponded during my entire time in service, had graduated from college and was now working in Lafayette, LA. We had 3 dates during my period of leave and then lost contact with each other.

I reported to Kelly Field in San Antonio as ordered to begin the process of separation from service. While there, I flew C-47s which are twin engine airplanes, the same as the DC3 passenger airplane. I chose to remain in the active reserves and was assigned to a reserve unit in New Orleans. I then received my terminal leave, the official date of separation from active duty being January 15, 1946. Subsequently I remained in the active reserves but had difficulty meeting the hours of training required of pilots for active reserve status because of the time demand of my medical studies. I was promoted to captain and on completing medical school I also had a military classification of Scientist-Special (Doctor). I did my flying at the New Orleans Lakefront Airport. We flew P-51s, AT-6s and AT-11s. As a civilian, while in North Carolina as a resident in Internal Medicine at Duke University Hospital, I flew an Aeronica Chief single engine plane.

Since my hours of flying lapsed due to my medical career, I was removed from military flying status on March 3, 1963. However I had and still have a civilian multi-engine pilot's license and certified instrument license. I have a total of 1044 flying hours.

I flew the following aircraft:

Piper J3	
PT- 23	open cockpit single engine
BT- 13	Vultee 450 hp single engine closed cockpit
AT 10	
B-17	Flying Fortress
P-38	
C-47	
P-51	
AT- 6	
AT-11	
Aeronica Chief single engine	

I received the following medals:

Good Conduct Medal
Air Medal with one Oak Leaf Cluster
Mediterranean Theatre Medal
European - African - Mediterranean Medal
American Theatre Campaign Medal
World War II Victory Medal

In addition our unit received 2 Presidential Unit Citations

I was discharged from the Air Force Reserve, on October 14, 1969, as a captain . In view of the fact that none of my crew suffered any injuries, I consider this an extraordinary experience and enjoyed every minute of it. I had an outstanding crew, each an expert in his field. They were all brave men of good character who carried out their duties willingly, with courage and expertise. Our crew had considerable camaraderie and had the "all for one, one for all" spirit. We had lots of fun and lots of close calls but these only further cemented our relationship. We kept in touch with each other although through the years gradually lost contact with some. Don Carrick, Al Dawson and I still keep in touch to this day.

A follow up on our crew:

Robert R. Burch   Pilot   New Orleans, LA
Went on to become a physician, now retired
Robert J. Moran   Co-pilot   Long Island, NY
Became Secretary of an International Banking Co., Deceased
Roger E. Sizer   Navigator   Milwaukee, WI
Lost tract of.
Arthur W. Fouts Jr.   Togglier   Florin, CA
Regional Sales Manager for Seagrams Co., Deceased
Archie B. Mills   Engineer   Ironto, VA
Lost tract of.
Donald Carrick   Radio Operator   Albermarle, NC
Remained in the active reserves, served in Korean War. Now retired from the Air Force
Alvin H. Dawson   Ball Turret Gunner   Sarasota, FL
Became President of a national brick company in Detroit, MI now retired in FL.
Nicholas J. Fotos   Waist Gunner   Annapolis, MD
Lost tract of.
Tim J. Holly   Waist Gunner   Chicago, IL
Lost tract of.
George H. Shirley Jr.   Tail Gunner   Los Banos, CA
Deceased

Snappie lived with the Carricks in Albermarle, NC and was a devoted pet. He lived to be 16 years and had to be euthanized because of advanced crippling arthritis.

Chapter 16
Photo Section

REGISTRATION CERTIFICATE

This is to certify that in accordance with the Selective Service Proclamation of the President of the United States

ROBERT (First name) RAY (Middle name) BURCH (Last name)

Lucy ST JOHN Louisiana (Place of residence)

(This will be identical with line 2 of the Registration Card)

has been duly registered this 30 day of JUNE, 1942

Albin J. St Pierre (Signature of registrar)

Registrar for Local Board ONE (Number) Lucy ST JOHN (City or county) LA (State)

THE LAW REQUIRES YOU TO HAVE THIS CARD IN YOUR PERSONAL POSSESSION AT ALL TIMES

D. S. S. Form 2 (Revised 6/9/41) 16—21631

(Registrant must sign here)

WW II draft registration card

BT-13 Basic Trainer

WAR DEPARTMENT
A. A. F. Form 121
Revised 1 October 1944

INDIVIDUAL A. A. F. ISSUE RECORD

NAME	GRADE
ROBERT R. BURCH	2d Lt.

RATING	ASN
1091	0833018

		QUANTITY ISSUED							QUANTITY TURNED IN						
		1	2	3	4	5	6	7	1	2	3	4	5	6	7
1. This form will be retained with the permanent records for the individual. 2. See A. A. F. Regulation 65-23 for items to be entered on this form. 3. The items to be listed on this form are nonexpendable and expendable-recoverable.	VOUCHER NO.			45-26P	45-227P				45-580-P	45-309	45-891-D	4			
	STOCK RECORD ACCOUNT NO.	Model Stock													
ARTICLES (Nomenclature, Type or Stock No.)	DATE	[illegible] 44	20 Dec 44	2/16/45	4/2/45	23 [illegible] 45			3/24/45	5-5-45	[illegible]	[illegible] 45			

Bag Ass'y B-4
Gloves Summer B-3
Goggles Ass'y B-8
Helmet, Summer ANH-15
Mask, Oxygen A-14
Suit, Summer OD ANS-31
Bag Flyers Kit A-3
Gloves Flying A-11
Helmet, Winter A-11
Jacket Intermediate B-10
Trousers Intermediate A-9
Shoes Winter A-6
Vest Life Preserver B-4
Kit Chute Emergency
Parachute Complete
Cylinder Bail Out H-1
Jacket, [illegible]
[illegible]
Glasses, [illegible], Flying
Watch, [illegible] SER. NO. 44-[illegible]7149
BAG-SLEEPING ARCTIC A-3
MATTRESS-PNEUMATIC
Pistol Cal 45
Holster M-3 Cal 45
Computor D-4
Protractor
Ammo Cal 45
Magazine

Individual's initials

Issuing officer's initials

(SEE REVERSE SIDE)

16—41708-1

Personal Combat Equipment Issued

Form 363A
(local)

Date 1/10/45

B-17G 44-8582

EQUIPMENT CHECK LIST

NOSE SECTION

ITEM	TYPE OR PART NO.	REQ'D	INST'D
Bomb Sight	M-9	~~1~~	0
Flexible Gun Sight	N-6A	1	1
Machine Gun, Caliber 50, Browning . . .	H39G5332	2	2
Cylinder, Oxygen, Walk Around	A-4	2	2
Drift Meter.	B-5	1	1
Fire Extinguisher	A-2	1	1
Astro Compass	MK-11	1	1
Machine Gun, Caliber 50, Browning . . .	H39G5332	2	2

TUNNEL SECTION

ITEM	TYPE OR PART NO.	REQ'D	INST'D
Rotary Inverter for Automatic Pilot. . . .	C-1	1	1
Servo Unit Motor	C-24860	1	1
Amplifier, A.F.C.E..	G-104	1	1
Control, Vertical Gyro	C-24858	1	1
Radio Compass Unit	BC-433	1	1
Cylinder, Oxygen, Low Pressure	G-1	7	7

PILOTS AND CO-PILOTS SECTION

ITEM	TYPE OR PART NO.	REQ'D	INST'D
Clock Assy., Pilots.	A-11	1	1
Turbo Boost Selector	G-1056A2CA2	1	1
Radio Control Box	BC-450	1	1
Radio Control Box	BC-451	1	1
Radio Control Box	BC-434	1	1
Radio Control Box	BC-602	1	1
Radio Control Box	BC-732	1	1
Pilots and Co-Pilots Cushion, Life Preserver	A-1	2	2
Pilots and Co-Pilots Cushion, Life Preserver	A-3	2	2
Pyrotechnic Pistol	P.S.-91638-2	1	1
Pilots Control Box A.F.C.E..	G1047C1CA1A2	1	1

UPPER TURRET SECTION

ITEM	TYPE OR PART NO.	REQ'D	INST'D
Cylinder, Oxygen, Walk Around	A-4	3	3
First Aid Kit		1	1
Inter Aircraft Lamp, Aladis.	C-3A	1	1
Fire Extinguisher	4-TB	1	1
Cylinder, Oxygen, Low Pressure	G-1	8	8
Machine Gun, Caliber 50, Browning	H39G5332	2	2

PAGE 1

Equipment Check List

MAIN CABIN SECTION

ITEM	TYPE OR PART NO.	REQ'D	INST'D
Dynamotor	PE-94	1	1
Headsets		10	10
Throatmikes.		10	10
Radio Receiver	BC-966	1	1
Cover, Pilots Enclosure.	15-9268-1	1	1
C6ver, Nose.	15-10614-86	1	1
Cover, Tail.	882657-2	1	1
Cover, Ball Turret	15-9009-1	~~1~~	0
Cover R.R. Hatch		1	1
Cover, Tail Gunner Window		1	1
Cover, Chin Turret		1	1
Cover, Engine		[illegible]	4
Armorers Tool Roll		1	1
Propellar Wrench Set		2	2
Cylinder, Oxygen, Walk Around	A-4	3	3
Machine Gun, Caliber 50, Browning	H39G5332	2	2
Reel Antenna	RL-42	1	1
Marker Beacon Receiver	BC-375	1	1
Motor, Servo Unit	G1004B	2	2
Kit, Radio Operators		~~2~~	0
Kit, Armorers		1	1
Kit, Crew Chief.		~~1~~	0
Auxiliary Power Unit with Extension. . . .	C-10	1	1

TAIL SECTION

ITEM	TYPE OR PART NO.	REQ'D	INST'D
Cylinder, Oxygen, Walk Around	A-4	1	1
Machine Gun, Caliber 50, Browning	H39G5332	2	2
Sight Flexible Gun	N-6A	1	1

OUTSIDE FUSELAGE SECTION

ITEM	TYPE OR PART NO.	REQ'D	INST'D
Machine Gun, Caliber 50, Browning	H39G5332	~~2~~	0 Radar
Sight Compensating Automatic	K-4	~~1~~	0

A. T. C. INSTALLED EQUIPMENT

ITEM	TYPE OR PART NO.	REQ'D	INST'D
Binoculars	M-3	2	2
Axe, Firemans Hand		1	1
Life Raft Accessory Kits		2	2
Dinghy Radio	SCR-578-A	1	1

COMPILED SHORTAGE LIST ON NEXT PAGE

Equipment Check List

UPPER TURRET SECTION CONTINUED

ITEM	TYPE OR PART NO.	REQ'D	INST'D
Automatic Compensating Gun Sight	K-3	1	1
Radio Receiver	BC-733	1	1

BOMB BAY SECTION

ITEM	TYPE OR PART NO.	REQ'D	INST'D
Bomb Release, RH and LH.	A-4	24	24
Bomb Shackel	B-10	24	24
Receptical, Bomb Release		42	42

RADIO COMPARTMENT SECTION

ITEM	TYPE OR PART NO.	REQ'D	INST'D
Transmitter.	BC-458	1	1
Transmitter.	BC-457	1	1
Transmitter.	BC-459	~~1~~	0
Transmitter.	BC-375	1	1
Receiver	BC-453	1	1
Receiver	BC-454	1	1
Receiver	BC-455	1	1
Receiver	BC-348	1	1
Dynamotor	DM-32	3	3
Antenna Relay Unit	BC-442	1	1
Modulator Unit	BC-456	1	1
Dynamotor	DM-33	1	1
Fire Extinguisher	A-2	1	1
Turbo Amplifier.	G-403A1	5	5
Dynamotor	PE-73	1	1
Reel Control Box	BC-461	1	1
Tuning Unit.	TU-26	1	1
Tuning Unit.	TU-5	1	1
Tuning Unit.	TU-7	1	1
Tuning Unit.	TU-9	1	1
Tuning Unit.	TU-6	1	1
Tuning Unit.	TU-8	1	1
Power Control Box	BC-958	1	1
Selector Control Box	BC-965	1	1
Cylinder, Oxygen, Walk Around	A-4	1	1
Frequency Meter, Complete	SCR-211	1	1
Hand Crank, Inertia Starter.		2	2
Crank Extension, Inertia Starter . . .		2	2
Cylinder, Oxygen, Low Pressure	G-1	3	3
Radio Receiver	BC-624	1	1
Radio Transmitter	BC-625	1	1
Radio Altimeter, Transmitter and Receiver .	RT-7	1	1
Antenna Tuning Unit.	BC-603-A	1	1
Machine Gun, Caliber 50, Browning . . .	H39G5332	1	1

Combat B-17G Equipment

In formation & all is quiet

Vapor trails on way to target

In formation

Flak over target

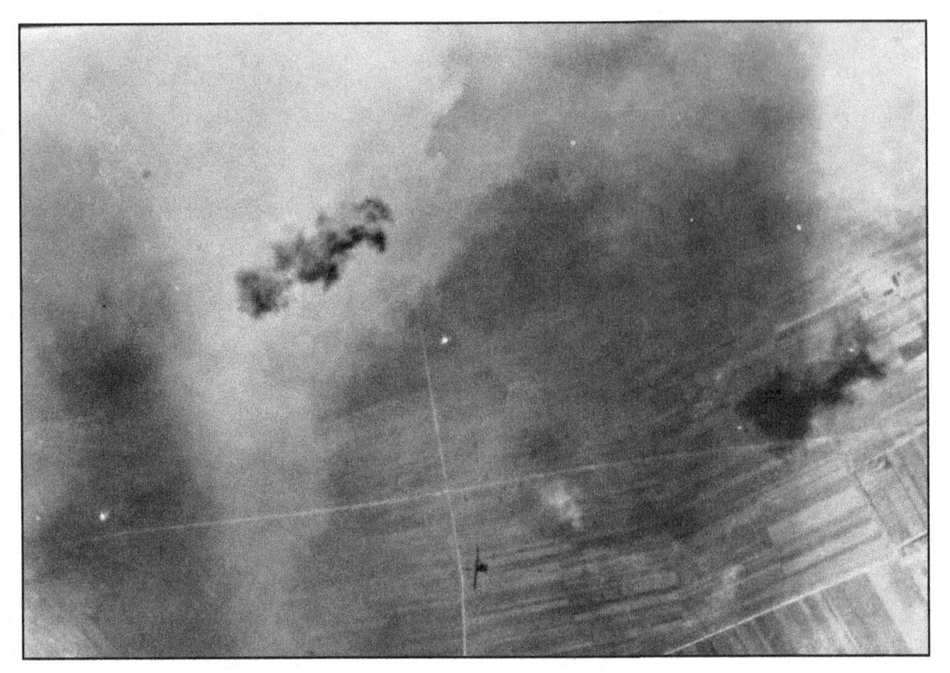

Plane hit and going down in pieces

Bombs away

Leaving a burning exploding target

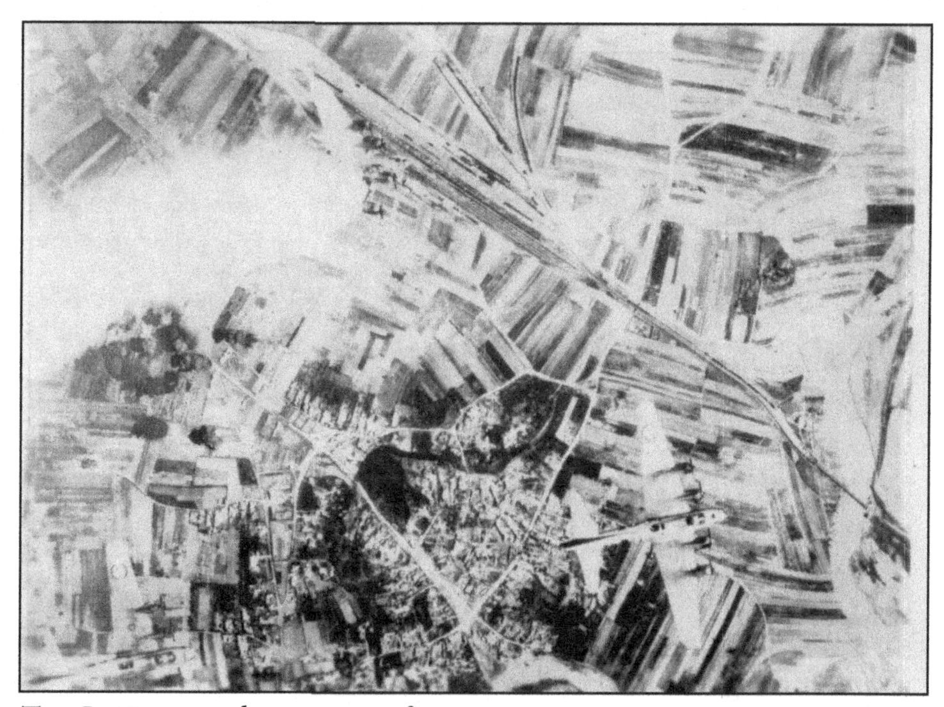

Two B-17s going down, one on fire

Number 3 prop feathered

Number 3 engine with an oil leak

Fouts, Mills, Burch and Sizer

Sizer, Burch & Jacoby
Our Tent

Al Dawson & Don Carrick L: 1945 R: 1990

Al Dawson, Tim Holly & Amos Healy

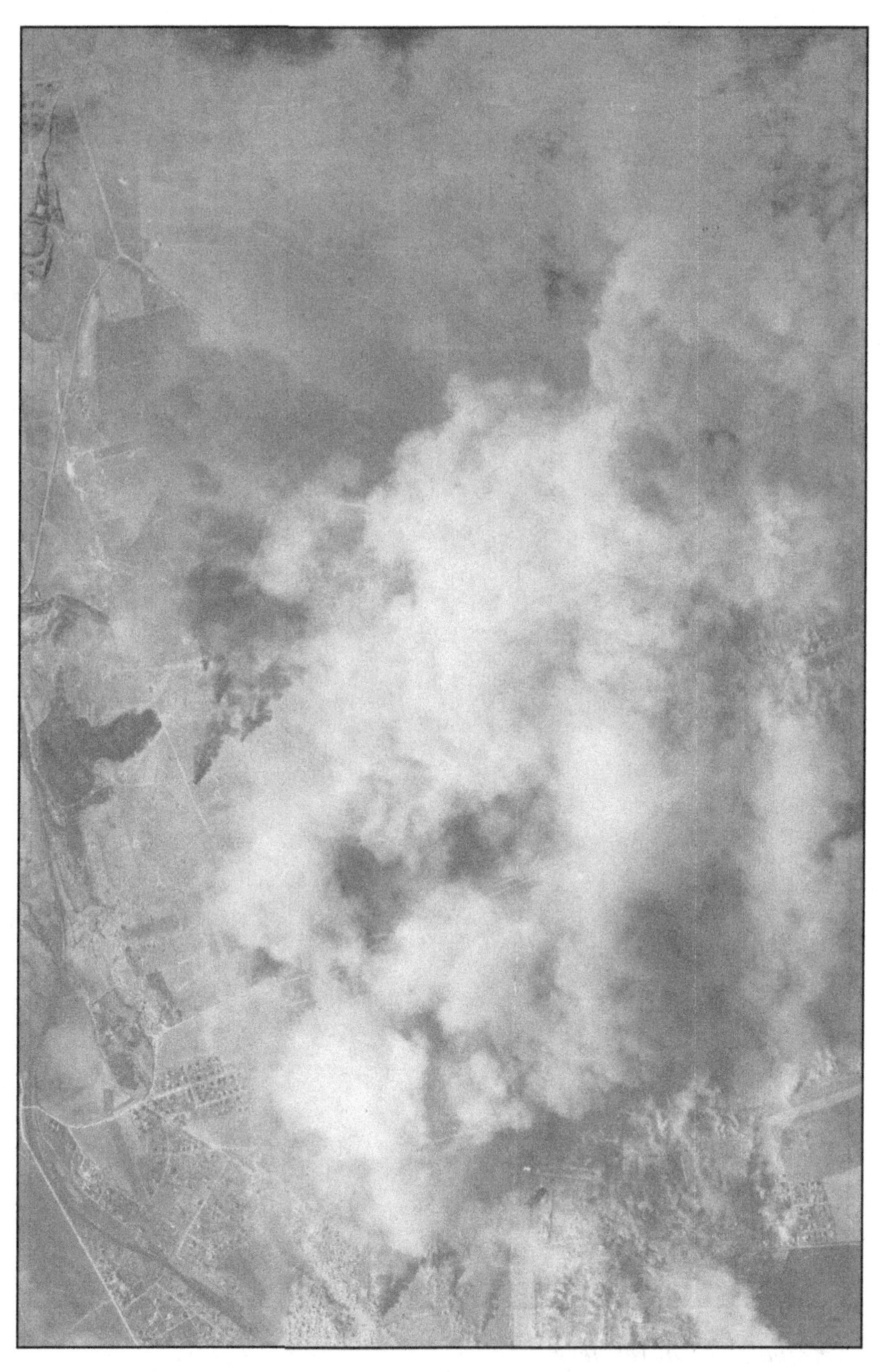

Prague, Czechoslovakia target

Verona, Italy oil refinery

George Shirley tail gunner in Marrakech, French Morocco

Snappie in North Carolina

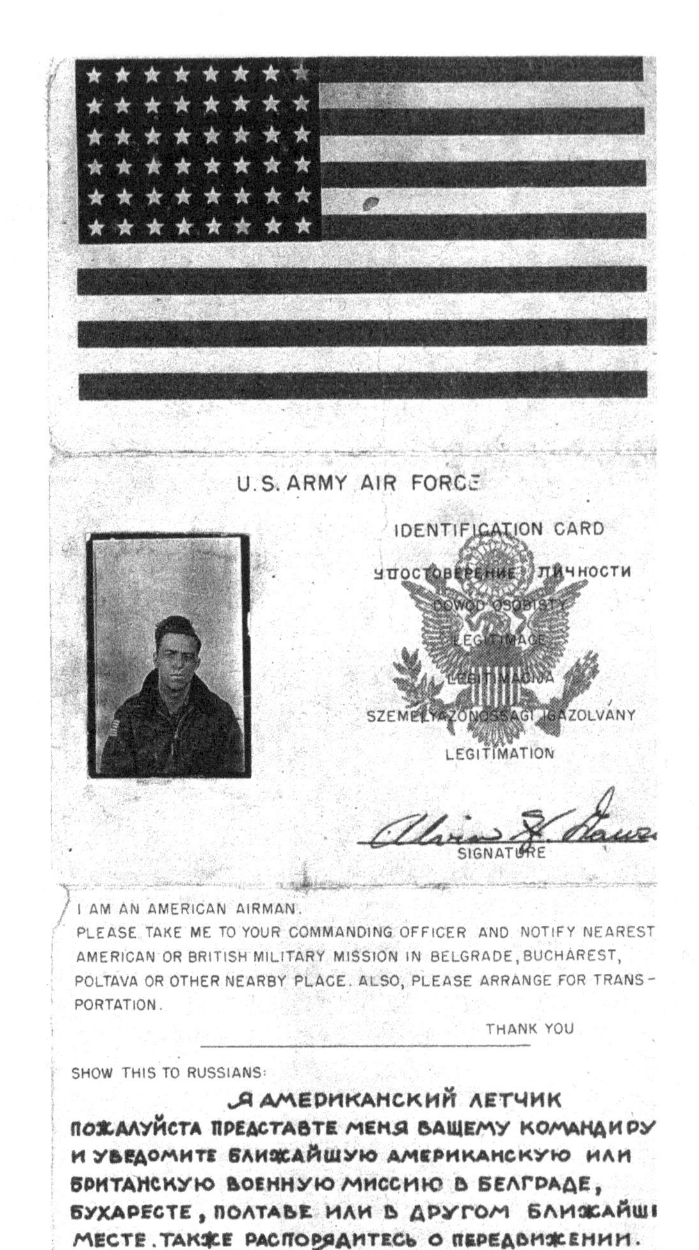

U.S. ARMY AIR FORCE

IDENTIFICATION CARD

УДОСТОВЕРЕНИЕ ЛИЧНОСТИ

DOWOD OSOBISTY

LEGITIMACE

LEGITIMACIJA

SZEMÉLYAZONOSSÁGI IGAZOLVÁNY

LEGITIMATION

SIGNATURE

I AM AN AMERICAN AIRMAN.
PLEASE TAKE ME TO YOUR COMMANDING OFFICER AND NOTIFY NEAREST AMERICAN OR BRITISH MILITARY MISSION IN BELGRADE, BUCHAREST, POLTAVA OR OTHER NEARBY PLACE. ALSO, PLEASE ARRANGE FOR TRANSPORTATION.

THANK YOU

SHOW THIS TO RUSSIANS:

Я АМЕРИКАНСКИЙ ЛЕТЧИК
ПОЖАЛУЙСТА ПРЕДСТАВТЕ МЕНЯ ВАЩЕМУ КОМАНДИРУ И УВЕДОМИТЕ БЛИЖАЙШУЮ АМЕРИКАНСКУЮ ИЛИ БРИТАНСКУЮ ВОЕННУЮ МИССИЮ В БЕЛГРАДЕ, БУХАРЕСТЕ, ПОЛТАВЕ ИЛИ В ДРУГОМ БЛИЖАЙШ МЕСТЕ. ТАКЖЕ РАСПОРЯДИТЕСЬ О ПЕРЕДВИЖЕНИИ.
БОЛЬШОЕ СПАСИБО!

Aviator's military I D (This is Dawson's)

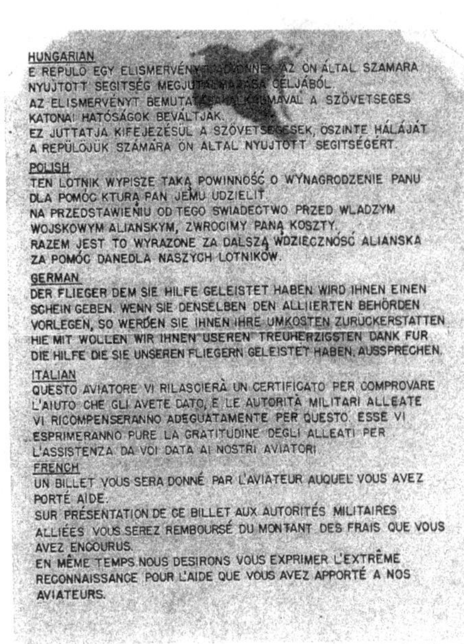

HUNGARIAN
E REPULO EGY ELISMERVÉN[illegible]NNEK AZ ON ÁLTAL SZAMARA
NYUJTOTT SEGITSÉG MEGJU[illegible]SA CÉLJÁBÓL.
AZ ELISMERVÉNYT BEMUTA[illegible]MAVAL A SZÖVETSEGES
KATONAI HATÓSÁGOK BEVÁLTJÁK.
EZ JUTTATJA KIFEJEZÉSÜL A SZÖVETS[illegible]SEK, ÖSZINTE HÁLÁJÁT
A REPÜLÖJÜK SZÁMÁRA ÖN ÁLTAL NYUJTOTT SEGITSÉGÉRT.

POLISH
TEN LOTNIK WYPISZE TAKĄ POWINNOŚĆ O WYNAGRODZENIE PANU
DLA POMÓC KTURĄ PAN JEMU UDZIELIT.
NA PRZEDSTAWIEŃIU OD TEGO SWIADECTWO PRZED WLADZYM
WOJSKOWYM ALIANSKYM, ZWROCIMY PANĄ KOSZTY.
RAZEM JEST TO WYRAZONE ZA DALSZĄ WDZIECZNÓSĆ ALIANSKA
ZA POMÓC DANEDLA NASZYCH LOTNIKÓW.

GERMAN
DER FLIEGER DEM SIE HILFE GELEISTET HABEN WIRD IHNEN EINEN
SCHEIN GEBEN. WENN SIE DENSELBEN DEN ALLIIERTEN BEHÖRDEN
VORLEGEN, SO WERDEN SIE IHNEN IHRE UMKOSTEN ZURÜCKERSTATTEN
HIE MIT WOLLEN WIR IHNEN USEREN TREUHERZIGSTEN DANK FUR
DIE HILFE DIE SIE UNSEREN FLIEGERN GELEISTET HABEN, AUSSPRECHEN.

ITALIAN
QUESTO AVIATORE VI RILASCIERÀ UN CERTIFICATO PER COMPROVARE
L'AIUTO CHE GLI AVETE DATO, E LE AUTORITÀ MILITARI ALLEATE
VI RICOMPENSERANNO ADEGUATAMENTE PER QUESTO. ESSE VI
ESPRIMERANNO PURE LA GRATITUDINE DEGLI ALLEATI PER
L'ASSISTENZA DA VOI DATA AI NOSTRI AVIATORI.

FRENCH
UN BILLET VOUS SERA DONNÉ PAR L'AVIATEUR AUQUEL VOUS AVEZ
PORTÉ AIDE.
SUR PRÉSENTATION DE CE BILLET AUX AUTORITÉS MILITAIRES
ALLIÉES VOUS SEREZ REMBOURSÉ DU MONTANT DES FRAIS QUE VOUS
AVEZ ENCOURUS.
EN MÊME TEMPS NOUS DESIRONS VOUS EXPRIMER L'EXTRÊME
RECONNAISSANCE POUR L'AIDE QUE VOUS AVEZ APPORTÉ A NOS
AVIATEURS.

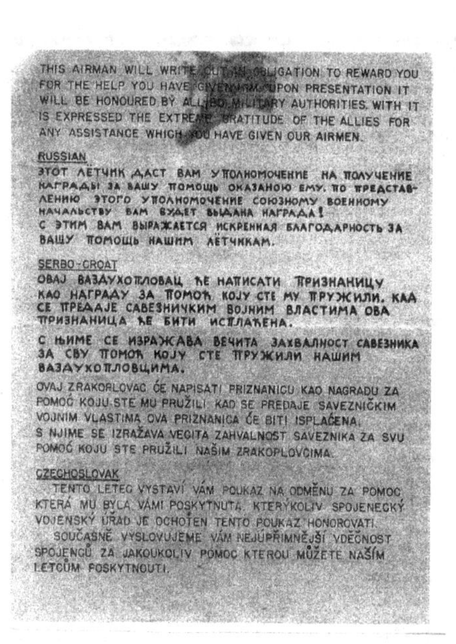

THIS AIRMAN WILL WRIT[illegible]IGATION TO REWARD YOU
FOR THE HELP YOU HAVE [illegible]UPON PRESENTATION IT
WILL BE HONOURED BY AL[illegible]RY AUTHORITIES. WITH IT
IS EXPRESSED THE EXTRE[illegible]RATITUDE OF THE ALLIES FOR
ANY ASSISTANCE WHICH [illegible] HAVE GIVEN OUR AIRMEN.

RUSSIAN
ЭТОТ ЛЁТЧИК ДАСТ ВАМ УПОЛНОМОЧЕНИЕ НА ПОЛУЧЕНИЕ
НАГРАДЫ ЗА ВАШУ ПОМОЩЬ ОКАЗАНОЮ ЕМУ. ПО ПРЕДСТАВ-
ЛЕНИЮ ЭТОГО УПОЛНОМОЧЕНИЕ СОЮЗНОМУ ВОЕННОМУ
НАЧАЛЬСТВУ ВАМ БУДЕТ ВЫДАНА НАГРАДА!
С ЭТИМ ВАМ ВЫРАЖАЕТСЯ ИСКРЕННАЯ БЛАГОДАРНОСТЬ ЗА
ВАШУ ПОМОЩЬ НАШИМ ЛЁТЧИКАМ.

SERBO-CROAT
ОВАЈ ВАЗДУХОПЛОВАЦ ЋЕ НАПИСАТИ ПРИЗНАНИЦУ
КАО НАГРАДУ ЗА ПОМОЋ КОЈУ СТЕ МУ ПРУЖИЛИ. КАД
СЕ ПРЕДАЈЕ САВЕЗНИЧКИМ ВОЈНИМ ВЛАСТИМА ОВА
ПРИЗНАНИЦА ЋЕ БИТИ ИСПЛАЋЕНА.
С ЊИМЕ СЕ ИЗРАЖАВА ВЕЧИТА ЗАХВАЛНОСТ САВЕЗНИКА
ЗА СВУ ПОМОЋ КОЈУ СТЕ ПРУЖИЛИ НАШИМ
ВАЗДУХОПЛОВЦИМА.

OVAJ ZRAKORLOVAC ĆE NAPISATI PRIZNANICU KAO NAGRADU ZA
POMOĆ KOJU STE MU PRUŽILI. KAD SE PREDAJE SAVEZNIČKIM
VOJNIM VLASTIMA OVA PRIZNANICA ĆE BITI ISPLAĆENA.
S NJIME SE IZRAŽAVA VEĆITA ZAHVALNOST SAVEZNIKA ZA SVU
POMOĆ KOJU STE PRUŽILI NAŠIM ZRAKOPLOVCIMA.

CZECHOSLOVAK
TENTO LETEC VYSTAVÍ VÁM POUKAZ NA ODMĚNU ZA POMOC
KTERÁ MU BYLA VÁMI POSKYTNUTA. KTERÝKOLIV SPOJENECKÝ
VOJENSKÝ ÚŘAD JE OCHOTEN TENTO POUKAZ HONOROVATI.
SOUČASNĚ VYSLOVUJEME VÁM NEJÚPŘIMNĚJŠÍ VDĚČNOST
SPOJENCŮ ZA JAKOUKOLIV POMOC KTEROU MŮŽETE NAŠÍM
LETCŮM POSKYTNOUTI.

Written in 9 languages, reward given for assistance to downed airman (Translation: This Airman will write out an obligation to reward you for the help you have given him. Upon presentation, it will be honored by Allied Authorities. With it is expressed the extreme gratitude of the Allies for any assistance given our Airman.)

15th. Air Force Patch

Print the complete address in plain letters in the panel below, and your return address in the space provided on the right. Use typewriter, dark ink, or dark pencil. Faint or small writing is not suitable for photographing.

(CENSOR'S STAMP) TO: Mrs. E. J. Wynnel, 2836 Eagle St., New Orleans, La. FROM: (Sender's complete address above) SEE INSTRUCTION NO. 2

GREETINGS FROM ITALY

DEAR ✓ Mother, Dad, Wife, Sweetheart, Sis ✓, Brother, Friend

HOW ARE YOU? ✓

I AM: Missing you ✓, Lonesome, Well ✓, Happy

WISH I HAD ✓: You, More sleep, Someone to love me ✓, Some kisses, More ambition

DOING LOTS OF ✓: Sight-seeing, Thinking of you, Work, Loafing, Sleeping, Nothing ✓, Movies, Everything, Traveling

HOPE YOU HAVE ✓: Everything you want, Behaved yourself, Been having fun ✓, Kept well, Been true to me, Not forgotten how to write ✓

ITALY IS ✓: Sunny, Beautiful, Hot, Smelly ✓, Picturesque

THE FOOD IS ✓: Fair ✓, Lousy, Good, Dehydrated ✓

ARE YOU ✓: Going out with 4F's ✓, Spending my allotment money, Staying out late at night, Missing me, Happy

THINGS ARE ✓: Fine, Rough, Slow ✓, Exciting

WILL BE ✓: Going to work, Thinking of you, Writing you again ✓, Going to bed

I HAVE SEEN ✓: U.S.O. Shows ✓, Fights, Baseball Games, Lots of good looking girls

XXX

YOURS ✓: Till the cows come home ✓, With love, Sincerely, Forever

NAME Bobby

AMERICAN RED CROSS - ITALY -

HAVE YOU FILLED IN COMPLETE ADDRESS AT TOP? REPLY BY V-MAIL HAVE YOU FILLED IN COMPLETE ADDRESS AT TOP?

Type letter written from combat to avoid bulk

ALLIED OFFICERS' CLUB

GIARDINO DEGLI ARANCI - TEL. DIAL 16298

Villanova - Napoli

MEMBERSHIP CARD

Member's Name Robert R Burck

S. A. MAGWOOD CAPT

Nº 96991 Treasurer SECRETARY

Membership Card to a club in Naples

Burch, Sizer & Jacoby

Don Carrick & Al Dawson

We can sail also
(L to R) Dawson, Sizer, Holly, Burch, Moran & Fouts on the Adriatic Sea

Allied Occupation Money

Famous WW II recruiting poster

Printed in the United States
by Baker & Taylor Publisher Services